Extrusion Coating, Lamination and Coextrusion

The complete process manual

B.H. Gregory

First published by Plastics Information Direct – a division
of Applied Market Information Ltd – in 2012
Republished by Elsevier

ELSEVIER

AMSTERDAM • BOSTON • HEIDELBERG • LONDON
NEW YORK • OXFORD • PARIS • SAN DIEGO
SAN FRANCISCO • SINGAPORE • SYDNEY • TOKYO

Extrusion Coating, Lamination and Coextrusion
The complete process manual

First published by Plastics Information Direct – a division of Applied Market Information Ltd – in 2012
Republished by Elsevier

Elsevier
Radarweg 29, PO Box 211, 1000 AE Amsterdam, Netherlands
The Boulevard, Langford Lane, Kidlington, Oxford OX5 1GB, UK
50 Hampshire St, 5th Floor, Cambridge, MA 02139, USA

ISBN: 978-1-906479-08-4

Contents

Introduction

This book is aimed at a wide readership of technologists, engineers, marketers and students engaged in the development, production or specification of multilayered products used in packaging and other industrial applications. The aim is to assist in optimizing product performance, evaluating the most cost effective solutions and providing useful information on the key polymers and films commercially available.

This book offers a practical and detailed evaluation of extrusion coating, coextrusion and film lamination processes using polyethylenes for the production of multilayered flexible materials. The technology and products are in constant development and it is, therefore, impossible to maintain a continuous up-date of every current innovation by this rapidly evolving industry. The body of work presented is intended to provide a comprehensive insight to extrusion coaters, laminators and converters of the various options that are available to produce tailor-made products and how to best exploit the available fabrication technologies. For the demanding process of extrusion coating this study focuses on aspects of extruder and die design, chill roll cooling and the critical optimization of adhesion to all substrates and films. Coextrusion, which is the complimentary process, is discussed in detail with the emphasis on melt stability, layer distribution, interlayer adhesion and physical properties. The study offers the laminator practical information on how to select, prepare and formulate films to maximize synergies. The wide palette of polymers, films, foils and pulp and paper products available are reviewed. All the adhesive materials and adhesion promotion methods are compared and will be of invaluable benefit to engineers and technologists working in this industry.

For the producer of packaging materials particularly in the sensitive perishable food and medical sectors, a number of attributes have to be embodied in the product according to the demands of the application.

Some are listed:

- Interlayer adhesion
- Heat sealability: no leakers
- Machineability: high-speed packaging
- Surface properties: easily printable
- Physical strength: impact and puncture resistance
- Balanced properties in the machine and transverse directions of the film
- Controlled moisture vapour transmission rate (MVTR)
- Controlled oxygen transmission rate (OTR)
- Product resistance: acids, alcohols, vegetable oils
- No loss of nutrients, aroma and moisture
- Heat sterilizable
- Opaqueness to UV light to reduce rancidity

These characteristics when specified must be delivered on high productivity machines with the most efficient use of materials to reduce waste and minimize carbon footprint. Dimensional control and overall product integrity must be perfectly matched to meet end-use specifications and the required shelf life.

The key processes that can integrate these properties either individually or in combination are:

- Extrusion coating
- Coextrusion
- Adhesive lamination

The extrusion coating process is an excellent method to combine a thin polymer layer, typically polyethylene with paper-based materials and aluminium foil. Coextrusion is used to combine a variety of different polymers that in combination will enhance properties such as gas barrier, heat seal and strength.

Lamination with reactive polyurethane adhesives is perhaps the most versatile since it can combine all flexible roll-fed materials made from the above two processes with others such as bi-oriented and metallized films to achieve the maximum benefits. All these processes will usually include at least one polyethylene layer to provide moisture protection and heat sealability and minimize costs.

Computer modelling

In this book much use is made of computer generated modelling to help in demonstrating and evaluating polymer flows in various parts of the processing equipment. This PC software was designed to simulate non-Newtonian flows in a variety of flow fields including screws, dies and coextrusion processes. The program solves the equations of conservation of mass momentum and energy for polymer melts through process equipment. The finite element method (FEM) is used, which is a rigorous solution of the differential equations. The Galerkin formulation is used to obtain solutions expressed as polynomials within small triangular elements and finally the individual solutions are assembled together to provide the velocity, pressure and temperature throughout the whole field of flow. From the velocity field the streamlines, local shear and stretch rates, stresses and particle residence times are calculated.

Computer programs are also used to predict heat transfer and cooling rates on chill rolls, air gaps and coextrusion dies.

The author believes the information discussed to be accurate and reliable. No liability is assumed or warranty given for the application of the data in this document. The data is based on technical information, which is available to the general public and to the best knowledge of the author it contains no confidential data received from a third party.

B.H. Gregory
2012

Chapter 1

The extrusion coating process

In extrusion coating a thin film of very hot molten polymer is pressed into a substrate passing through a cooled nip roll assembly as shown in **Figure 1.1**. Coating thickness will normally range from 10-75 g/m^2. For fabric and textile coating, thickness may be above 500 g/m^2.

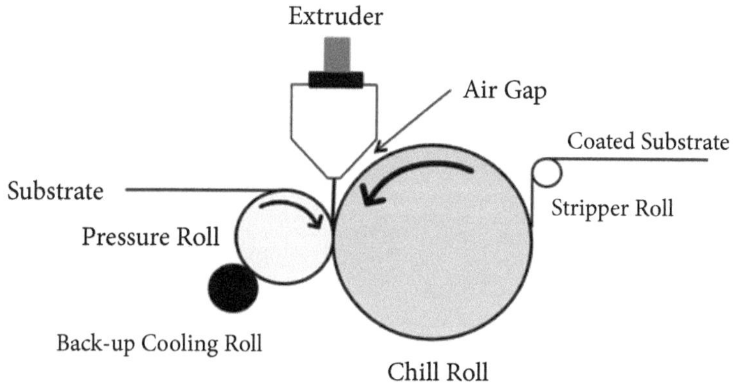

Figure 1.1

Basic extrusion coater

The section of the equipment that combines the polymer melt with the substrate(s) is sometimes referred to as the 'laminator.' Usually the term 'coater' is more descriptive of the function to avoid confusion with the adhesive lamination process described in chapter 7. The polymer melt at uniform temperature flows from the lips of the slot die and is drawn down into the nip formed by the chill roll and the rubber sleeved pressure roll. The substrate is simultaneously fed into the same nip where it combines with the molten polymer under pressure. A compression force is generated in the nip, which forces the melt to penetrate the substrate. The cooled web is then separated from the chill roll as it passes over a stripper roll and is pulled by the downstream haul-off equipment at constant tension. The positioning of the extruder relative to the chill roll is critical. The extruder should be mounted on a hydraulically operated carriage, which allows movement in three directions. The melt entry angle and height (air gap) of the die relative to the nip is critical. The extruder should be equipped with a screenpack unit to filter any contamination from the melt.

In extrusion **laminating** two substrates fed from rolls are combined with the polymer melt acting as the adhesive between the two webs. **Figure 1.2** shows the extrusion lamination process. This high-speed technique is widely used to laminate aluminium foil to paper with low density polyethylene (LDPE).

The coating section is the heart of the machine where the melt has to spread and adhere to the substrate(s). In the nip the melt must be fluid enough to flow and penetrate into the moving surface. The polymer side in contact with the chill roll will freeze immediately as a temperature gradient rapidly develops across its interface. The bond must be formed virtually instantaneously. Optimum nip pressure is required to maximize adhesion between the melt and the substrate. A pneumatic pressure of 9-18 kg/linear cm is a useful starting point. Excessive pressure will not improve adhesion and may cause unnecessary stress on the roll bearings. The nip force will not influence the thickness of the coating, which is determined by the output of the extruder relative to the rotational velocity of the chill roll. A water cooled (15-20 °C) steel back-up roll removes excess heat from the surface of the rubber pressure roll as shown in **Figures 1.1** and **1.2.**

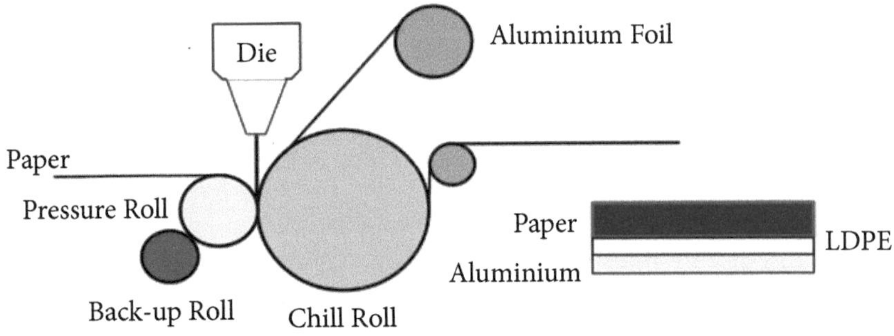

Figure 1.2

Extrusion lamination

Figure 1.3 shows a schematic cross section of a typical extrusion coating line. The substrate delivered at constant tension from the unwind arrives into the coating nip via a series of idler rolls. A steam heated roll is a useful option to improve adhesion by preheating/drying the paper substrate before it arrives in the nip. The surface of the substrate will usually be treated by either corona or flame to maximize adhesion. The corona discharge electrodes should be positioned as **close** as possible to the melt stream exiting the die.

The melt enters into the nip almost tangential to the chill roll. An overlap of 3-5 mm on the pressure roll prior to entry in the nip is recommended. The air gap or distance between the die and nip will have a profound influence on adhesion.

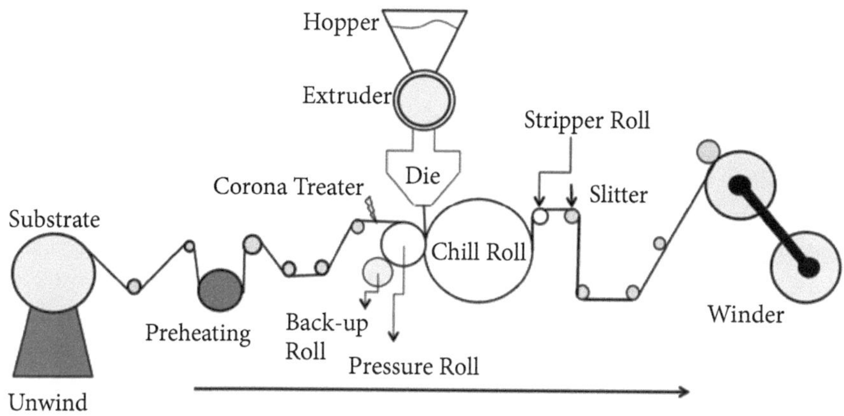

Figure 1.3

Schematic cross section of a coating line

A useful starting distance is 150-200 mm (6-8″) of air gap. However further adjustments will be necessary as experience with different polymers and operating conditions are developed. The effect of time in the air gap on the polymer's **oxidation** and its influence on adhesion is critical and is comprehensively discussed in this study.

The coated product is stripped off the chill roll by a set of driven nip rolls, which are synchronized with the drives for the chill roll and winder to accurately control web tension. The practical recommended arrangement is to position the stripper roll so that the web comes off the chill roll along a vertical line from its horizontal axis. Increasing the overlap beyond this point does not seem to improve chill roll release. As will be shown later in this chapter the chill roll temperature, its diameter, line speed and surface finish are crucial in determining the operational efficiency of the machine.

The excess polymer at the edges (overhang) is cut by the slitter knives positioned past the chill roll where web tension is controlled and held constant. The slitting can be by razor, score or shear cutting. The straight edged web is taken up by a turret or surface winder where rolls of predetermined size are collected. Commercial machines incorporate automatic (on the fly) splicing, which allows uninterrupted roll changes at both ends of the process. A typical commercial process will run in the range of 150-250 mpm (metres per minute) line speed. Higher capacity machines have been constructed.

Extruder sizes will vary from 114 mm (4½″) to 150 mm (6″) diameter depending on the required output relative to the die width and line speed specified. Larger machines are offered. Extruders with polyethylene outputs of 1000 kg/hr. are in operation.

Typical die widths range from 1.2 to 3.5 m for most paper and board coating applications. Dies of 4.0 m width have been built for fabric/textile coating. Coextrusion, usually based on combining feedblock designs, is widely used to include thin adhesive or barrier layers in the structure. These satellite extruders will vary in size from 50-90 mm (2″-3½″) diameter and must be optimally sized to minimize residence time and avoid melt stagnation with very thin layers of polymers sometimes of relatively low heat stability. For some applications dual manifold designs where the melts combine near or just outside the die exit are preferred.

Drool catchers must be provided to protect the machine from the melt curtain as the extruder(s) carriage is driven to and from the coater.

To operate this complex process sophisticated computerized management systems are required. The following key process variables must be monitored and controlled:

- Extruder output(s)
- Melt temperature(s)
- Automatic dies with feedback thickness control
- Polymer feed(s) (silos, boxes etc.)
- Substrate tension/feed control(s)
- Blenders
- Treatment (corona, flame)
- Edge trim
- Automatic splicing and roll changes
- Labeling of finished rolls

1.1 Chill roll cooling section

The coater in its most basic form is made up of a rotating steel chill roll and a hard rubber sleeved pressure roll, which form the nip, plus a stripper roll to feed the cooled coated web to the downstream haul-off equipment. The hot polymer melt is drawn onto the substrate by passing it through the nip where a pneumatic load is applied. The nip pressure loading should be precision controlled to ensure repeatable impressions across the chill roll.

The melt flow rate at the entry to the nip determines the coating thickness as the melt spreads across the chill roll. In the nip the cold roll exerts a compression force, which squeezes the melt into the surface of the substrate, usually paper, which is porous. The melt will bond to the surface by a combination of mechanical penetration and various physical-chemical bonding forces. The complex subject of adhesion with different substrates and polymers is discussed in chapter 5.

The efficiency of the cooling section of the machine is a key factor in determining productivity. The chill roll is essentially a heat exchange unit that cools the hot polymer to ambient temperature and enables the transfer of the coated web to the haul-off equipment with a minimum of pulling (stripping) force from the chill roll.

Good design and maintenance are essential to hold a uniform temperature over the total surface of the chill roll. Double walled spiral rolls are utilized and designed to keep the temperature variation across the face to ±1-2 °C. A heavy inner shell supports a thin outer shell with spiralled flights for maximum heat transfer efficiency. The chill roll should be designed for rapid changeover to enable the use of different surface finishes. Rotary unions through which the cooling water flows should also be designed for quick disconnects and the roll change should be made within half an hour.

Figure 1.4 illustrates the flow of coolant into the chill roll via a rotary union.

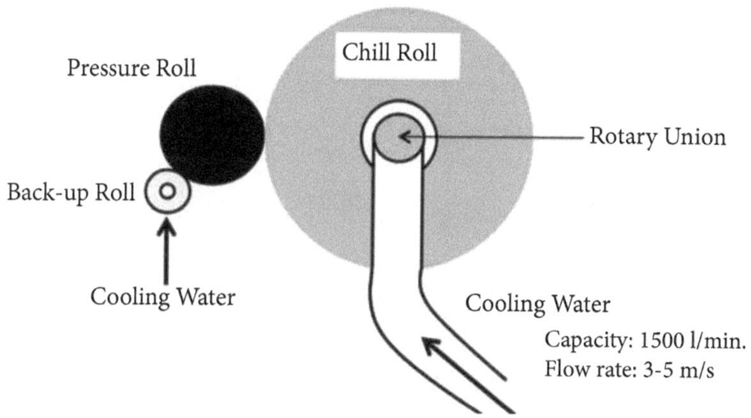

Figure 1.4

Chill roll cooling

Chilled water at 15-20 °C is pumped through the roll. The surface temperature should be higher than the environmental dew point of water to avoid condensation. The most efficient chill rolls use a very thin outer steel shell with a special assembly procedure to prevent the shell collapsing. High conductivity steel must always be used in the construction. Clean soft water must be used to keep the inner channels clean. The water capacity on a commercial line should be around 1500 l/min. at a flow rate of 3-5 m/s. Periodic flushing, cleaning and de-scaling is essential to maintain the system at maximum heat transfer efficiency.

The back-up cooling roll is also fed with water via a similar method.

Refrigerated water is necessary on high-speed lines. If the chill roll temperature is too high the coated polymer will stick to its surface and may tear from the substrate as it is pulled away. However, heated chill rolls are sometimes used in specialized applications to maximize adhesion with specific materials. Finally, too low a temperature may decrease adhesion due to excessively rapid quenching of the melt.

The effectiveness of the chill roll cooling process was studied by computer modelling. The basic model is shown schematically in **Figure 1.5**. The process conditions are as shown below.

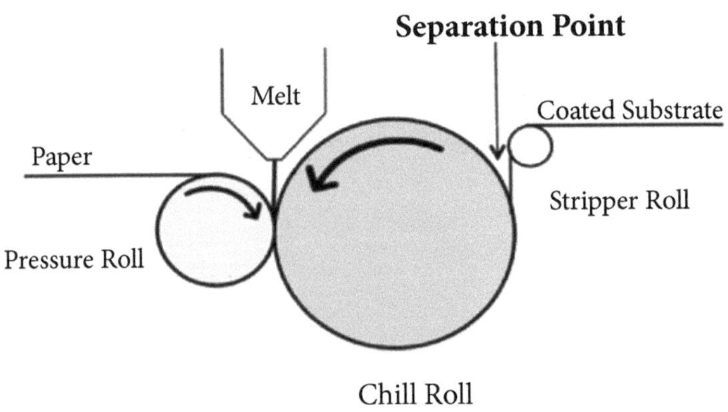

Figure 1.5

Chill roll set-up

- Chill roll diameter: 450 mm
- Pressure roll diameter: 250 mm
- Chill roll temperature: 18 °C
- Pressure roll surface temperature: 40 °C
- Width of rolls (web): 1500 mm
- LDPE temperature at nip: 315 °C
- Paper: 0.15 mm, temp. 35 °C

- Mass flow rate (LDPE): 500 kg/hr.
- Line speed: 200 mpm
- Coating weight: ~27 g/m^2
- Ambient air temperature: 20 °C

The first step was to calculate the heat exchange that occurs at the nip where the hot polymer (LDPE) is pressed against the paper and starts cooling rapidly as it contacts the chill roll set at 18 °C. **Figure 1.6** shows the cross cut where the temperature profile is plotted.

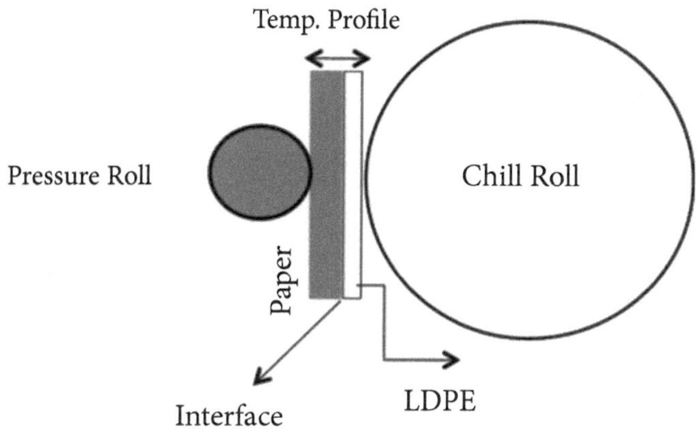

Figure 1.6

Cross cut at paper-LDPE interface

Figure 1.7 shows the temperature profile across the LDPE-paper interface as it passes through the **nip**. The LDPE temperature drops instantaneously from 315 to 235 °C. The substrate (paper) enters the nip at 35 °C: it is assumed that some heat is transferred from the corona/flame treater and the hot die. The temperature gradient through this section is critical since it will determine the wetting and penetration of the melt into the substrate's surface.

Further downstream at the **separation** (stripping) point from the chill roll the **interface** temperature has dropped to ~21.5 °C as shown in **Figure 1.8**.

The simulation illustrates the efficiency of this cooling process as 500 kg/hr. of LDPE is drawn at a line speed of 200 mpm by the machine. Over a distance of half a revolution of the 450 mm diameter roll the melt temperature drops from 315 to approximately 21 °C. The coating is now sufficiently cooled to be released from the roll by the haul-off equipment. The optimum release temperature will depend on the surface finish of the roll and the characteristics of the polymer. From **Figure 1.8** the coated web has a temperature ranging from 21-23 °C at an ambient temperature of 20 °C. This virtual experiment demonstrates the very rapid heat exchange from the melt to the chill roll over this relatively short distance of travel.

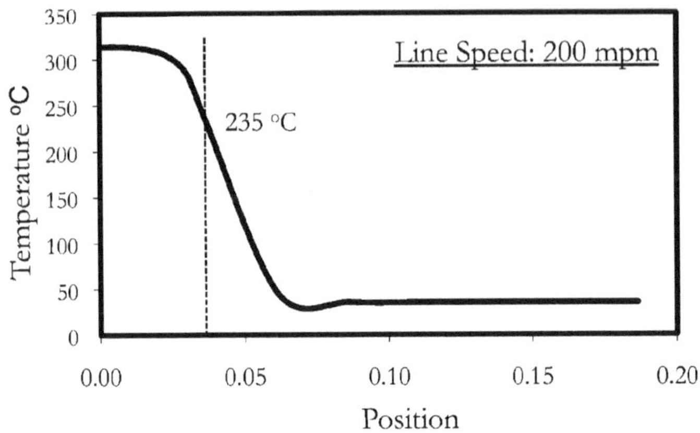

Figure 1.7

Temperature profile of cross cut at nip

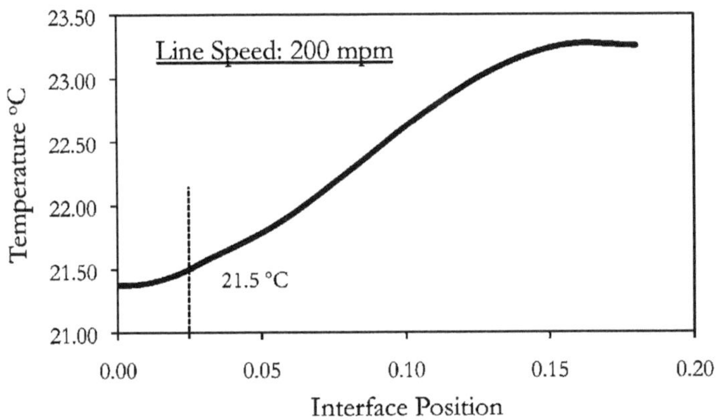

Figure 1.8

Temperature at separation point from chill roll

Using the same process conditions and model the effect of varying the chill roll temperature from 10 to 24 °C was evaluated. The results are in **Figure 1.9**. At 10 °C the release temperature drops to 14 °C and the value increases as shown with increasing chill roll temperature. With the chill roll at 24 °C the interface temperature is 27 °C. Smooth release with LDPE is now unlikely. Deciding on the optimum temperature setting will be based on achieving excellent release and maximizing adhesion. Too rapid cooling may compromise the ability of the hot melt to interact with the substrate and will reduce melt stress relaxation time. The result is weakened adhesion.

In **Figure 1.10** the effectiveness of three chill roll diameters are compared. The LDPE at 315 °C was fed at 1000 kg/hr. and line speed was 400 mpm. The coating weight was 27 g/m² as in the previous simulations. Chill roll temperature was set at **18 °C** in all cases. The cross cuts were taken at the **separation** point from the chill rolls.

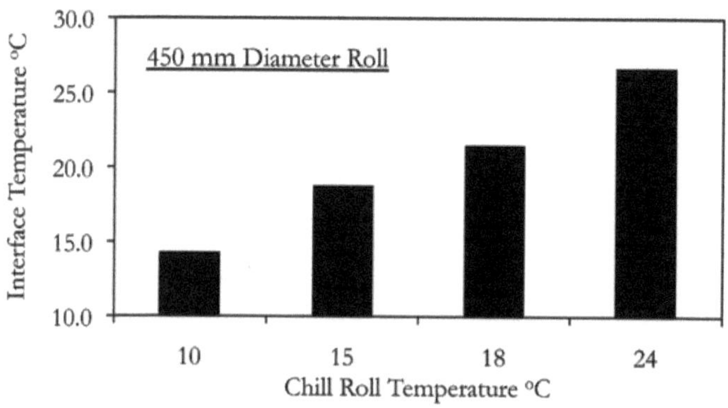

Figure 1.9

Cooling at 4 chill roll temperatures

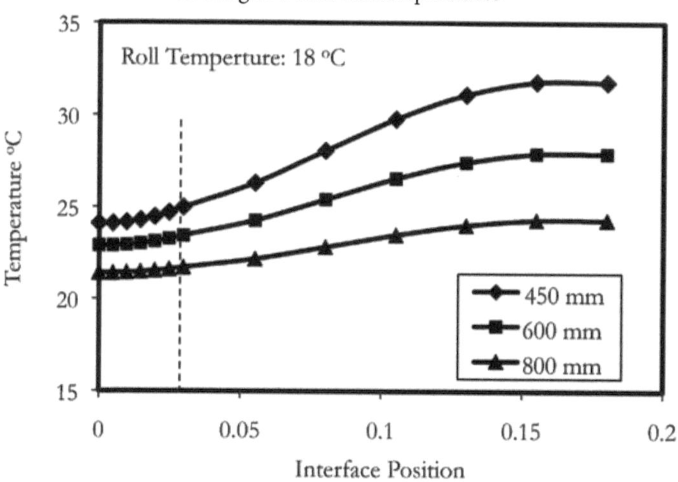

Figure 1.10

Temperature at cross cut for 3 chill roll sizes

The interface temperatures show a difference of approximately 4 °C (21-25 °C) between the three chill rolls at the separation zone. These simulations are at extreme conditions operating at a line speed of 400 mpm. At these conditions the largest of the three chill rolls is preferred.

In **Table 1.1**, the 450 mm diameter roll is run at four temperature settings and two throughputs. Doubling the line speed to 400 mpm increases the release temperature by some 3-4 °C for all temperature settings. Unless run at 15 °C or lower this 450 mm diameter roll is not recommended for high-speed operation.

Table 1.1

Cooling at various temperatures and 2 outputs (450 mm diameter roll)

Chill roll temp. (°C)	500 kg/hr. (200 mpm)	1000 kg/hr. (400 mpm)
10	14.3	17.9
15	18.8	***21.8***
18	***21.5***	25.0
24	26.7	29.0

If it is assumed that the web release temperature with LDPE should be in the range of 20-22 °C the data in **Table 1.2** shows that the 800 mm diameter roll will perform well at 18 °C whereas the smaller roll needs a temperature of 15 °C.

Table 1.2

Relationship between chill roll diameter and cooling

Diameter	Chill Roll	Separation
450 mm	15 °C	21.8 °C
800 mm	18 °C	21.7 °C

Over time the higher cost of the 800 mm diameter roll will be off-set by the cost of extra refrigeration capacity to drop the surface temperature to 15 °C with the smaller chill roll.

The limitation of these simulations is that ideal conditions are assumed where the contact is perfect and the chill roll surface temperature is unchanged. It is crucial that the cooling capacity of the plant maintains a uniform temperature across the roll surface throughout the production run. It is also important to note that although the polymer coating temperature has dropped to around 21 °C as it leaves the chill roll the substrate (paper) may be relatively warm as it absorbs heat from the melt and from the surface treatment process, such as corona or flame treatment. This is approximately the average temperature at which the web may reach the winder. As the coated paper roll cools to a lower ambient temperature during storage crushing forces may develop in the bulk roll as the web shrinks. These may result in unwinding difficulties in the downstream converting process.

These calculations are intended to illustrate the interdependence of chill roll size with surface temperature in relation to the throughput of the machine. Establishing the optimum release temperature from a given chill roll surface finish is clearly critical for efficient machine operation.

Finally, the polymer may also play an important role. As an example if HDPE (high density polyethylene) is blended with LDPE (low density polyethylene) release is very significantly improved allowing for a higher chill roll temperature. Conversely with EVA copolymers (ethylene-vinyl acetate) chill roll sticking can be problematic unless specially designed roll surfaces are used.

However, these virtual experiments demonstrate the outstanding efficiency of the chill roll cooling process in enabling very high production rates with melts delivered at very high temperatures.

1.2 Chill roll surfaces

The chill roll surface must be carefully selected as it will influence the finish of the coating in the following ways:

- Release from chill roll surface
- Optical properties
- Slip or coefficient of friction (COF)

Chrome plating is recommended for all chill rolls to minimize wear.

Matte chill rolls are the most commonly used particularly with pulp and paper substrates. These have a peak to valley surface finish of usually 10-40 μm and produce surfaces with low coefficients of friction (COF).

The gloss chill rolls have a peak to valley surface of 1-2 μm. The resulting coating with LDPE is very glossy and transparent. However, the coating is difficult to strip away from the roll particularly at elevated speeds. Sticking can be severe enough to cause delamination of the coating from the substrate. COF is also very high and electrostatic surface dusting with finely divided starch powder will be required if the material is to be run on form, fill and seal packaging machines (this is not desirable). Since most extrusion coating substrates are opaque (e.g. paper/board and aluminium foil) transparency is usually not an issue.

N.B. Polypropylene (PP) can produce very high gloss coatings without roll sticking.

The mirror pocket finish was developed as a compromise solution to achieve high gloss and low coefficient of friction. This is essentially a chrome plated finish interspersed with fine undulations that are placed on the surface and then polished to remove any roughness. **Figure 1.11** illustrates schematically the difference between a matte and mirror pocket finish. The results in **Table 1.3** show some typical gloss and COF values for an LDPE coated on kraft paper with three different roll finishes.

Table 1.3

Effect of roll finish on film surface properties

Chill roll	COF	% Gloss
Gloss	0.5	12
Matte	0.3	1
Mirror Pocket	0.4	10

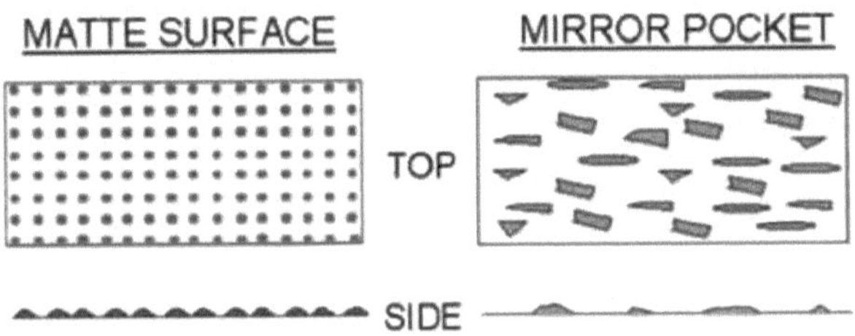

Figure 1.11

Two types of chill roll surface finish

In specialized applications such as thermal lamination films, which use **adhesive** polymers, embossed chill rolls are used. To avoid sticking the chill roll is acid solution etched to a pattern with a depth of 50-150 μm. The surface is optionally sandblasted to 80-100 RMS (root mean square) and usually chrome plated. The adhesive material such as EVA copolymer (16-18 % VA) is extrusion coated on the substrate, typically 12 μm OPET (oriented polyester) film.

Most coaters will have chill rolls with various finishes to meet different application requirements.

1.3 Pressure roll

The pressure steel roll is fitted with a sleeve made from a heat and abrasion resistant rubber composition with a Durometer hardness of at least 80 Shore A. Soft rolls will increase the contact surface at the nip, but pressure will be lower. Recommended roll diameters and hardness for different materials are shown in **Table 1.4.**

The sleeves are usually made from either chloroprene or Hypalon rubber (DuPont de Nemours). Silicone rubber has also been used because of its excellent release properties. However, silicone elastomers have poor abrasion resistance and cannot easily be formulated to a Shore hardness >80 A.

Table 1.4

Pressure roll specifications

Substrate	Roll diameter (mm)	Hardness (Shore A)
Paper/paperboard	350	85-90
Plastic films, glassine, cellophane	180	90-95

Finally, the back-up cooling roll should be of double shell construction for maximum efficiency. This will allow for higher outputs and optimum nip load adjustment.

It is important that the edges of the pressure roll be protected from the hot polymer overhang. Otherwise, the polymer will stick to its surface causing the overhang at the edges to split away from the coated substrate as well as ultimately damaging the rubber sleeve. The best method of protection is to use reinforced Teflon endless belts (5-10 cm wide per side), which lap the pressure roll at the contact area of the overhang. The belts should also wrap around the cooling back-up roll to remove excess heat. An example of an endless belt set-up is shown in **Figure 1.12**. The hot polymer overhang is applied on the reinforced belt from which it is easily stripped away. The belts are either fibreglass or Kevlar impregnated. In order to combat heat distortion problems from thick coating edges chilled air is blown on the tape.

Past the chill roll edge trim slitters will be setup. **Figure 1.13** shows three slitting techniques used on extrusion coating machines. Razor slitting is carried out by placing the knives right at the edges of the web held under controlled tension between two rolls. Knives are placed in series to ensure that as the first knife is blunted the next one is in-line for trimming. Groove slitting includes grooves on the roll into which the knife penetrates. Several grooves are included to adjust for varying web widths. In score cutting a rotating knife slits the edges over a groove.

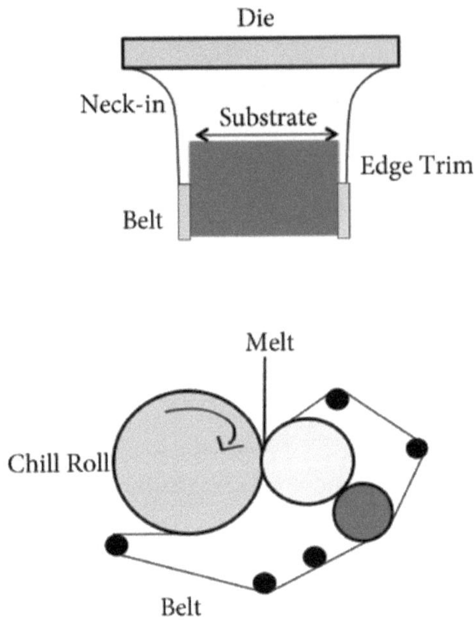

Figure 1.12

Endless Teflon belt set-up for edge trim

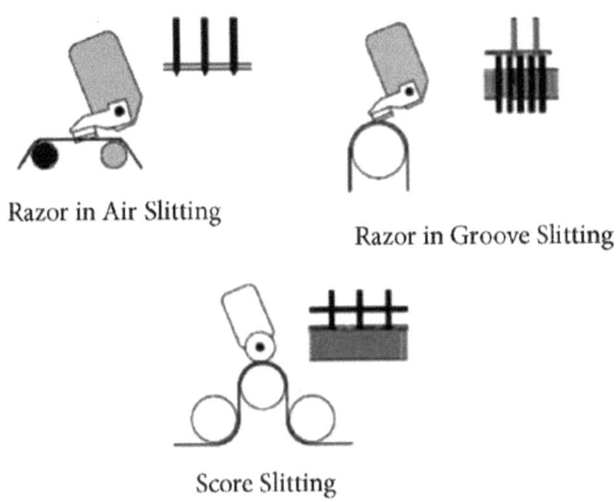

Figure 1.13

Some edge slitting techniques

1.4 Web tension control

The objective of any web handling system is to be able to deliver flat rolls of finished material free of wrinkles and creases. Perfect alignment and precise tension control of web handling equipment is a prerequisite when specifying an extrusion coating line. If tension is too low the result will be rolls with baggy edges, wrinkles and telescoping caused by slippage. Excessive tension will also cause wrinkles as localised stresses build-up in the web. Corrective devices such as, spreading rolls (banana shaped) and side guiding rolls may be necessary in some situations.

The forces acting on a rotating roll are illustrated in **Figure 1.14**. The torque is the turning force on a rotating device and is calculated as the product of the tension and the radius of the rotating shaft.

$$\tau = \omega \times r \qquad (1.1)$$

τ = torque
ω = tension
r = radius

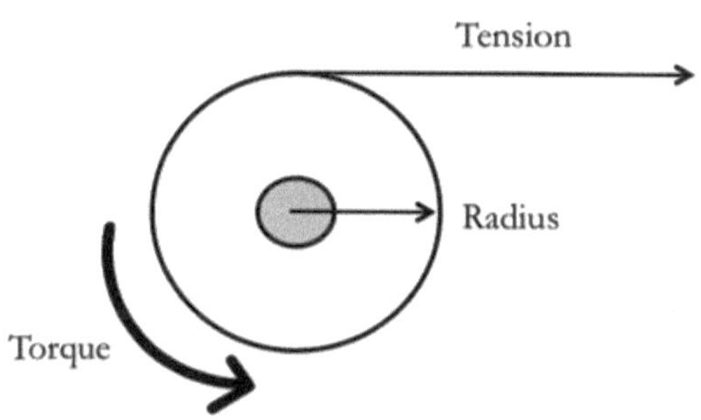

Figure 1.14

Forces acting on a roll

The total tension is the linear tension or stress applied to pull the web multiplied by the web width. Some recommended web tensions for the most commonly used substrates and films are listed in **Table 1.5**. Excessive web tension can cause wrinkling, stretch or in extreme cases web break. Low tension will also result in wrinkles.

Closely regulated DC driven solid state tension control systems should be provided. The drives must be synchronised for precise speed control match with the other positively driven units of the machine. If the machine is intended to handle a wide range of products then clearly the cost and complexity of the web handling system will escalate proportionally.

Table 1.5
Web tensions for typical substrates

Substrate	Tension (N/cm)
Paper (g/m²)	
25	0.89
33	1.34
50	1.79
65	2.68
100	3.57
130	4.47
Paperboard (µm)	
460	5.36
305	7.14
380	8.93
510	12.50
635	16.07
763	19.65
Films/Foil	
Al foil (9 µm)	1.79
BOPP (25 µm)	0.89
BOPP (50 µm)	1.79
Cellophane (18 µm)	0.89
Cellophane (25 µm)	1.34
LDPE (25 µm)	1.34
LDPE (50 µm)	2.68
OPET (12 µm)	0.45
OPET (25 µm)	0.89
PA (18 µm)	0.27
PA (25 µm)	0.45

1.5 Winders and splicing

Roll and core handling systems ranging from manual to fully automatic are available from machinery constructors. These enable high-speed on-the-fly splicing without interruptions. Machines capable of handling roll width of 2.5-3.0 metres weighing up to 500 kg have been designed. Line speeds of 600 mpm are claimed.

The function of the winder is to produce flat and straight edged rolls. The roll should be hard to the touch across its width. Winding equipment with inadequate tension control will produce telescoped rolls or other deformities such as hills and valleys on the surface. However, even with well-designed winders deformed rolls can still occur. One of the principle culprits is the carryover of air by the web into the roll. Layers of air are carried on both sides of a fast moving web and enter the roll as it rotates. This causes uneven tension leading to deformed rolls.

Surface winders as shown in **Figure 1.15** are better able to block the air and stop it from entering the roll. The pressure of the roll against the driven drum winder eliminates air entrapment. Tension is accurately controlled with an adjustable speed motor that drives the drum or surface winder. The roll itself rotates by surface friction. The surface velocity of the roll being formed is equivalent to that of the drum. To avoid slippage, pressure is applied by a pair of pneumatically driven arms, which force the roll against the rotating drum.

With a turret or centre winder as shown in **Figure 1.16**, the roll is formed on a rotating shaft. Tension is controlled by the rotational velocity of the shaft, which is propelled by an adjustable speed motor. As the roll radius grows its surface velocity would increase with the drive at a constant rotational speed. Thus for a constant web speed the motor driving the core or shaft must progressively slow

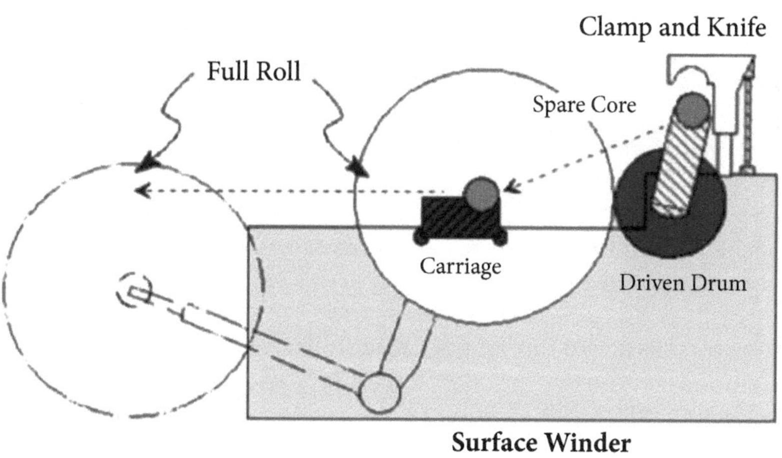

Figure 1.15

Schematic of surface winder

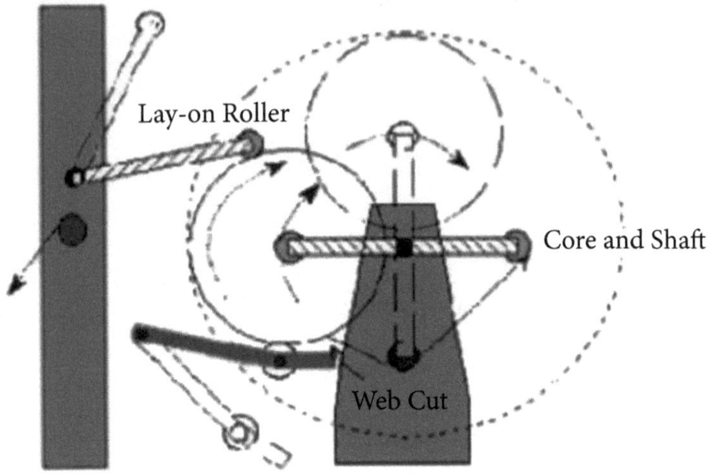

Figure 1.16

Schematic of centre winder

down. In **Figure 1.14**, the torque equation shows that as radius increases the torque on the drive will increase at constant tension. As a result tension control becomes difficult with this system, since as the roll diameter changes the torque on the shaft must be continuously adjusted to maintain constant tension. Therefore, with centre winders it is very important to include high quality tension control systems if rolls free of creases and telescoping are to be produced.

In addition, to prevent air from entering the roll a lay-on roller should be installed. The arms of the lay-on roller are pneumatically activated and force it against the roll. This forms a nip that excludes air from entering. The pneumatic pressure is varied to suit the required roll hardness.

Both types of winders are designed with on the fly splicing facilities and the best are capable of executing roll changes at speeds of up to 500 mpm with paper and paperboard substrates.

There are several ways to carryout automatic roll changes. Widely used methods are bump and cut, and the stationary knife. The bump and cut method is shown in **Figure 1.17**. This method is used at lower line speeds (<250 mpm) and is the simplest. An adhesive tape is placed on the fresh core as the finished roll is moved to its final position. The bump roll is pneumatically actuated and the knife is released. The knife splices the web, which adheres to the tape on the core and winding is continued. The system creates fold back and requires tape.

Figure 1.18 shows the stationary knife method, which can be operated at higher line speeds and has no fold back. The cut off knives are on the turret and not the roll changer. The fresh core is moved to the change position, which is around 15 cm from the lay-on roll. The pneumatically driven knife is actuated moving it close to the core. The turret is indexed onto the lay-on roll and the knife moves into the path of the web. The knife is not slammed into the web. This proprietary design ensures a well wound roll with the elimination of waste at changeover.

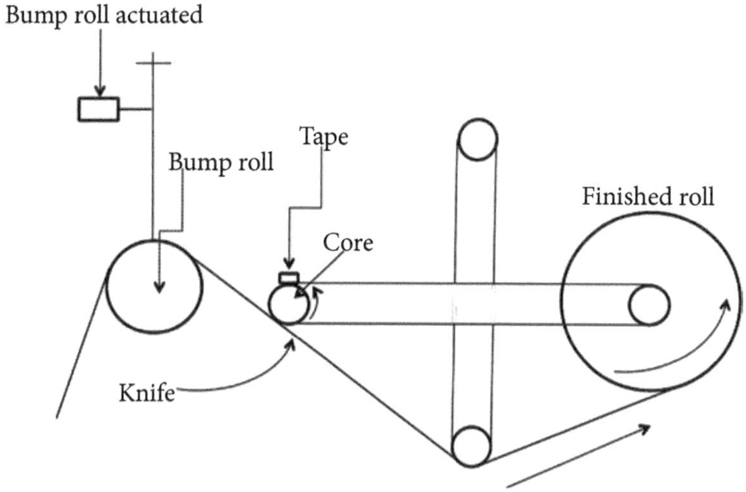

Figure 1.17

Bump and cut roll change

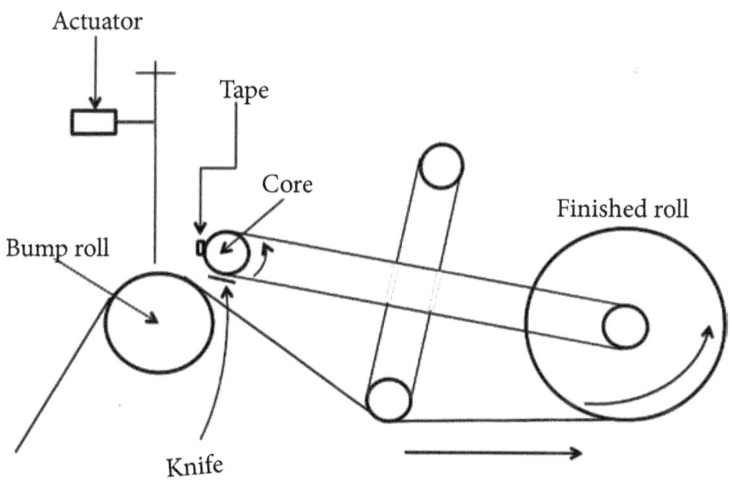

Figure 1.18

Stationary knife method

Finally the use of shafted versus non-shafted rolls should be considered. Heavy core steel shafts can lead to operator fatigue. On very wide lines aluminium may not be strong enough. Carbon fiber and composite lightweight shafts are available and designed with automatic shaft inflation, which removes the chance of the operator neglecting to inflate the shaft.

Shaftless unwind operation removes the extra handling and facilitates roll loading. Normally these systems do not allow widely different widths to be run on each spindle. However, new designs can include adjustable arms and make it possible to accommodate different widths.

1.6 Stretching flows and neck-in

A critical part of the process is the drawdown or stretching of the melt from the die exit to the chill roll nip. **Figure 1.19** illustrates this process. The melt at high temperature is stretched in the gap at a high rate of deformation and must be able to resist the applied forces without breaking at the set process conditions. The edges will neck-in as shown in **Figure 1.19**. These edge effects are generated by a combination of forces, which include extrudate swell, edge stress and surface tension. The result is edge thickening usually referred to as edge beading and is sketched in **Figure 1.20**. The die width must be adjusted with deckles so that all the thicker material overhangs beyond the substrate's edges and is trimmed by the slitter knives further downstream.

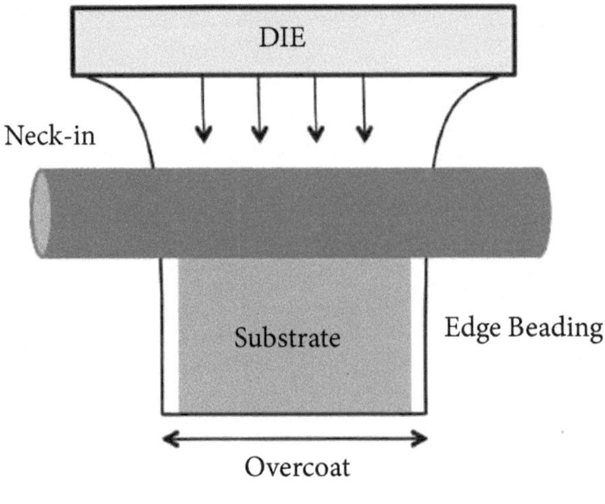

Figure 1.19

Neck-in and drawdown

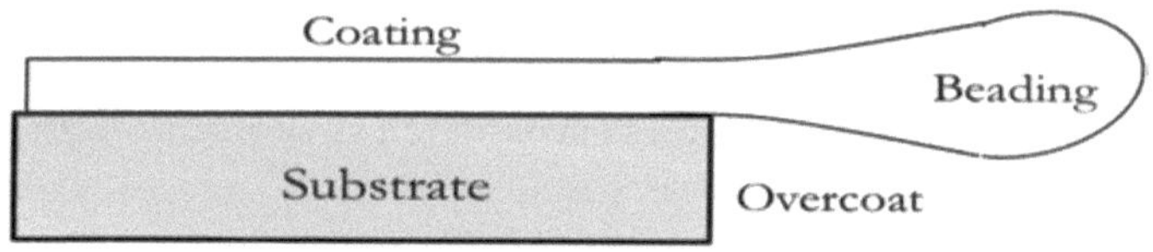

Figure 1.20

Edge beading

The edge trim can account for approximately 5% of throughput and must be minimized. Edge trim in extrusion coating is normally discarded due to the risk of contamination if the highly oxidized polymer is recycled in the process. In some instances, such as drool material, this may be collected in blocks and compounded and granulated by specialist companies for reselling as substandard material for non-critical applications.

Figure 1.21 shows schematically two forms of stretching flows. The most common is the strain thinning response and is typical of the linear polyolefins, polyesters and polyamides. The tension hardening behaviour is exhibited by high pressure LDPE and copolymers such as EVAs (ethylene-vinyl acetate) and EAAs (ethylene-acrylic acid). The ringed part of the curve is within the elongation rate range encountered in the extrusion coating process. Neck-in is controlled by the strain hardening element of the elongational flow response. Therefore, as strain hardening increases with strain rate, neck-in is reduced with increasing line speed.

On the downside high elongational viscosity will lead to high stresses, which may result in brittle failure. Consequently LDPE **cannot** be drawn into very thin

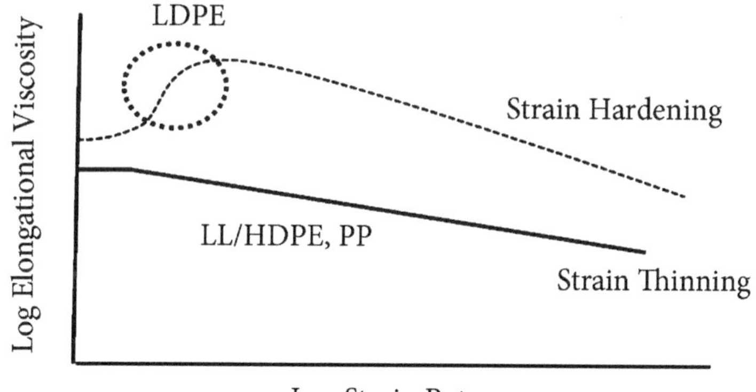

Figure 1.21

Strain hardening and thinning materials

films or fibres. In contrast strain thinning HDPE and PP are fibre forming and can be stretched at high draw ratios to form biaxially oriented films. However, at the draw rates and melt temperatures encountered in extrusion coating, LDPEs designed for this process can achieve coating weights and line speeds which easily exceed the requirements of existing commercial applications.

Elongational or extensional viscosity is difficult to measure experimentally. Although the Trouton ratio (1.2) for Newtonian fluids rigorously correlates elongational viscosity (η_e) with shear viscosity (η) in uniaxial extension, it is unfortunately invalid for non-Newtonian flows.

$$\text{Trouton ratio: } \eta_e = 3\eta \qquad (1.2)$$

In the case of non-Newtonian melts the function η_e can increase or decrease with strain rate. Data derived from shear flow measurements is, therefore, inadequate for predictive capacity when considering stretching flows. In this study the Trouton relationship has been adapted and is used to help clarify the changes taking place in the air gap as the melt is deformed at high velocity. The calculated velocity and strain rate data are valid, but the elongational stress values are fictitious and used here only for comparative purposes.

The response of the LDPE melt to stretching in the air gap was studied with computer modelling. The process conditions entered are in **Table 1.6** for a typical flat die. The calculated velocity in the 160 mm air gap is shown in **Figure 1.22**. The estimated average LDPE thickness drawn from the 0.6 mm die gap is 24 μm.

The ratio between the velocities at the die lips and nip roll is 1:72. Velocity at the lips can be calculated with equation 1.3.

$$Q = V \times t \times W \times \rho \qquad (1.3)$$

Q (kg/s) = mass flow rate
V (m/s) = velocity
t (m)= die gap
W (m) = die width
Melt density (ρ) = 750-760 kg/m^3

The calculation demonstrates the rapid rate of dimensional deformation that takes place in the air gap. The stresses generated will be defined by the elongational viscosity of the polymer, draw rate and melt temperature. **Figure 1.23** shows the effect of line speed on neck-in for two polymers. In the case of the LDPE neck-in is reduced. With polypropylene (PP) neck-in is increased. These two polymers clearly respond differently to extensional stress as shown in **Figure 1.23**.

Table 1.6

Flat die process model

Die width	1500 mm
Die gap	0.6 mm
LDPE flow rate	500 kg/hr.
Melt temperature	310 °C
Air temperature (air gap)	50 °C
Line speed	180 mpm
Air gap	160 mm

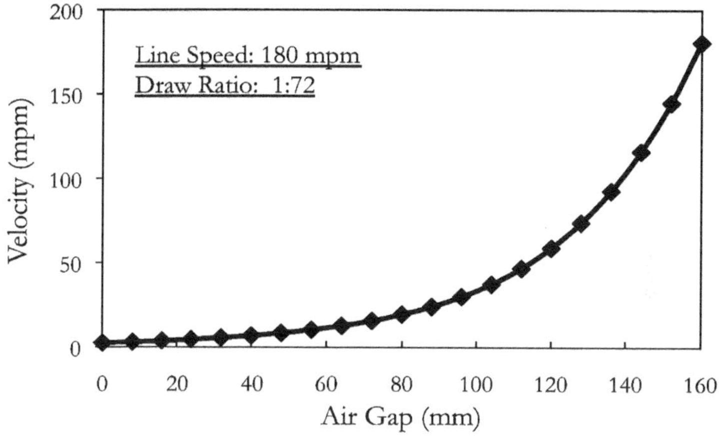

Figure 1.22

Velocity in 160 mm air gap

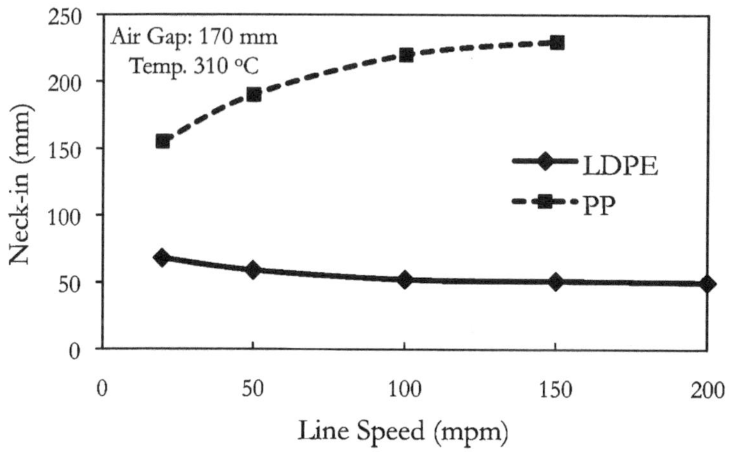

Figure 1.23

Neck-in of LDPE and PP

Elongational stress applied on the melt can be moderated by the following machine parameters:

- Die gap (die lip opening)
- Melt temperature (viscosity)
- Line speed (velocity)
- Air gap

Figure 1.24 plots the prediction of elongational stress versus air gap length at **two** die gap settings. The stress data as discussed earlier is fictitious. However, the differences between the two curves are of interest. **Figure 1.24** shows that the wider die gap results in higher stress in the melt as it approaches the nip. The effect is to reduce neck-in. This suggests that polymers with low extensional viscosity such as polypropylene should be run with wider die gaps to minimize neck-in. The effect as can be seen is not substantial.

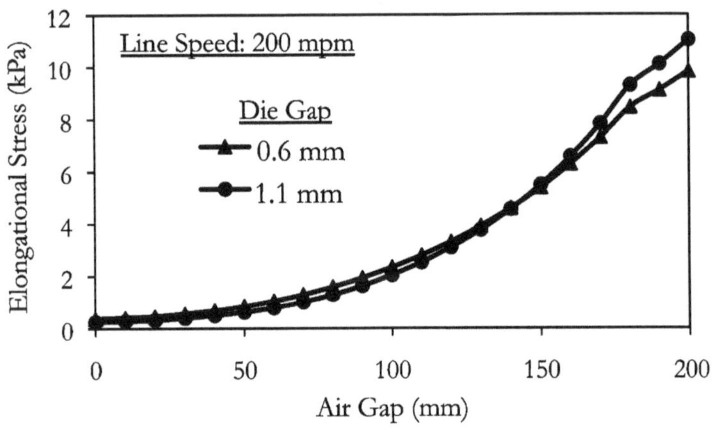

Figure 1.24

Elongational stress at 2 die gaps

Wide die gaps will reduce die pressure and will compromise melt distribution across the die. For LDPE die gaps of 0.5-0.6 mm are suggested and wider gaps can be experimented with for other materials. Wide die gaps may in fact be necessary with materials that develop excessive pressures in the die.

Figure 1.25 shows the effect of melt temperature on neck-in. Neck-in is reduced with decreasing temperature and increasing line speed. At 275 and 295 °C the melt breaks above 100 mpm where the applied elongational stress exceeds the melt strength.

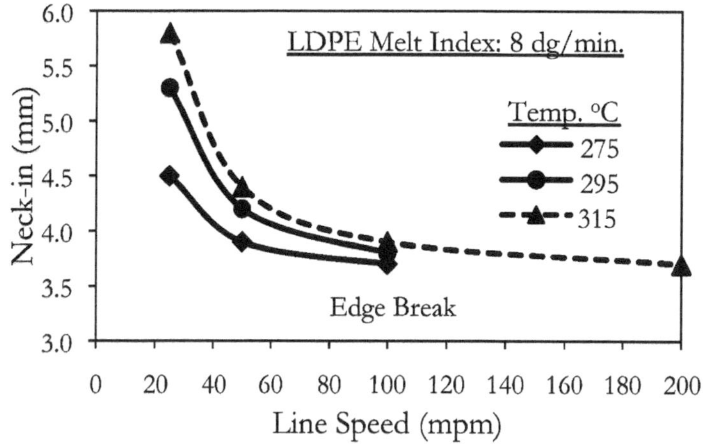

Figure 1.25

Effect of temperature on neck-in

A corollary to these observations suggest that the ethylene copolymers (EVA, EAA etc.), which are extruded at some 30-40 °C lower temperature will exhibit lower neck-in than processes running with LDPE at higher temperatures.

The effect of air gap on elongational stress was estimated and shown in **Figure 1.26**. The calculations were at constant line speed and temperature. The graph clearly demonstrates that elongational stress is reduced very significantly as the air gap is extended. The 10 mm gap is typical of the cast film process and

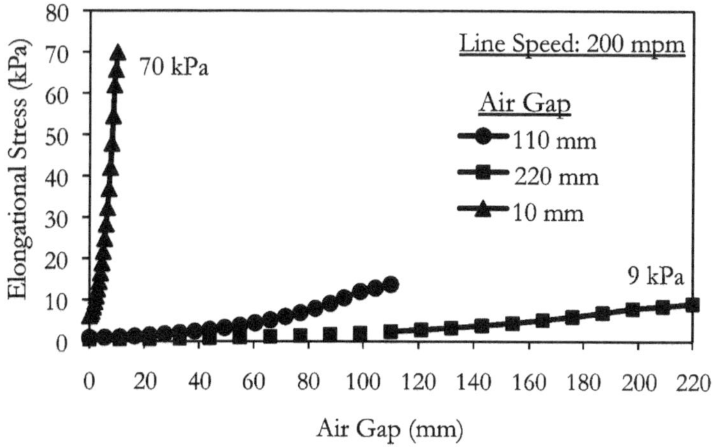

Figure 1.26

Elongational stress at 3 air gaps

shows that there is approximately a tenfold drop in elongational stress as the air gap is increased to 220 mm. The calculated stress values are approximate, but the differences in the values are relevant. The conclusion is to use short air gaps when processing tension thinning polymers, such as polypropylene. This has a much more substantial effect on neck-in than enlarging the die gap (**Figure 1.24**).

Finally the effect of sourcing the LDPE from different manufacturers was evaluated.

A group of LDPEs normally used for extrusion coating are compared for neck-in in **Figure 1.27**. All have a melt index of ~8 dg/min. The air gap was 150 mm and the melt temperature 315 °C.

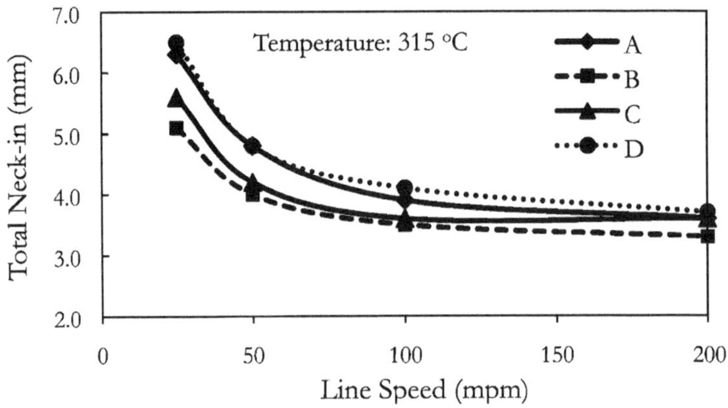

Figure 1.27

Neck-in of 4 LDPE grades

The data shows the influence of the LDPE grade on neck-in. Clearly grade B is the best and D the worst. Another observation is the flattening of all the curves beyond 100 mpm. The neck-in drops significantly from 25 to 100 mpm, but beyond this velocity the neck-in response is minimal. It appears that the stresses developed in both machine and cross directions reach a maximum level that are capable of resisting the forces causing the melt to neck-in. Excessive line speed will cause melt break.

In **Figure 1.28** an interesting study by Oveby[1] shows that coating grade LDPEs of similar melt index can vary considerably in drawdown. From **Figure 1.28** drawdown can vary by as much as 200-400 mpm to break. The polymerization

1 Oveby C. Rheological Indicators of LDPE to Predict Processing Performance in Extrusion Coating. Annual Transactions of the Nordic Rheology Society, vol. 13, 2005.

conditions and reactor design will control the molecular architecture of the polymer, which will affect its viscoelasticity and elongational viscosity. Judicious LDPE grade selection by the coater can help to minimize neck-in and edge trim.

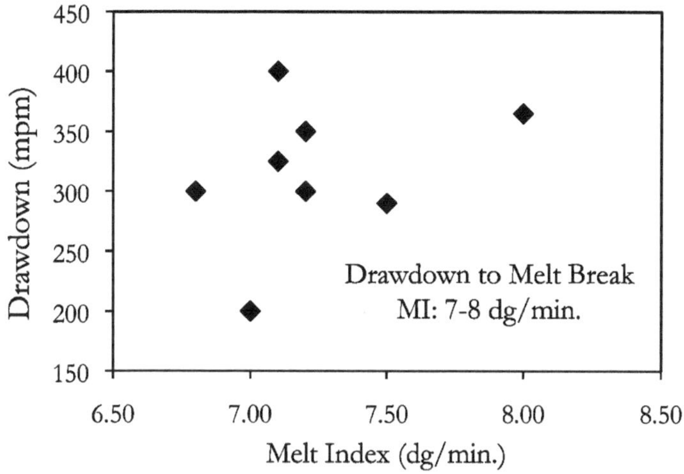

Figure 1.28

Drawdown variation at constant melt index
Data reproduced with permission from Oveby C.[1]

1.7 Draw resonance

Draw resonance occurs with:

- LLDPE (linear low density polyethylene)
- HDPE (high density polyethylene)
- PP (polypropylene)
- PA (polyamide)
- PET (polyester)

The onset of draw resonance is unmistakable and described as follows.

> *'Sustained oscillation or pulsation of the polymer melt with respect to the velocity and cross-sectional area of a melt drawing process. The phenomenon occurs between the die and the take-off position at a fixed velocity at the die and a fixed velocity at the take-off position. It manifests itself as irregularities in the final coating dimensions and often produces widely variable thickness. When line speeds significantly exceed the speed of onset, draw resonance can cause web breaks'.*

The only solution to avoid draw resonance is to reduce line speed (elongation rate) until the phenomenon disappears. Wider die gaps can also help, but the effect is minimal (**Figure 1.24**). Blending PP or HDPE with LDPE will retard the onset of draw resonance. HDPE and LDPE pellets can be easily dry blended in a tumbler drum or hopper blender prior to extrusion. The same applies to LLDPE blends. Blending with PP is more difficult and requires an additional homogenization process. The polyamides and polyesters are incompatible with all polyethylenes.

Chapter 2

Extrusion coated products

Typical extrusion coating applications and machine die widths are listed in **Table 2.1**. Commercial line speeds vary from 150 to 300 mpm. Faster machines have been constructed. The feasibility of 1000 mpm lines has been evaluated. For many of the commodity applications, such as multiwall sacks, paper bags and sugar pouches the basic single station line is the preferred machine. For more complex structures the 2-station tandem line as shown in **Figure 2.1** is an extremely versatile process. The tandem machine enables two side coating in a single pass plus lamination from roll-fed substrates. With the addition of coextrusion the capability of the machine is significantly extended. The web handling equipment must have variable tension control capabilities to handle different substrates. Some recommendations are given in **Table 2.2**.

Table 2.1
Typical extrusion coating applications

Process	Application	Die width (mm)
Paper	Multiwall sacks, bags, sugar pouch, photo-paper, roll wrap, ream wrap, cement bags, cable wrap, release paper base, thermal films.	1500-3000
Flexible packaging	Lidding, wrappers, snack food, soups, condiments, medical packaging, toothpaste, toiletries, glassine, shampoo.	500-2000
Paper (card)board	Aseptic milk/juice cartons, paper cups, paperboard, boxes.	1200-3500
Fabric scrim/raffia/ non-wovens	Sacks, hospital drapes, medical gowns, diapers, tarpaulins, breathable laminates, military packaging.	2000-4000

Table 2.2
Web tension recommendations

Material	Weight / thickness	Tension (N/cm)
Paperboard	300 (g/m^2)	7.0
Paper	35 (g/m^2)	1.8
OPET (film)	12 μm	0.45

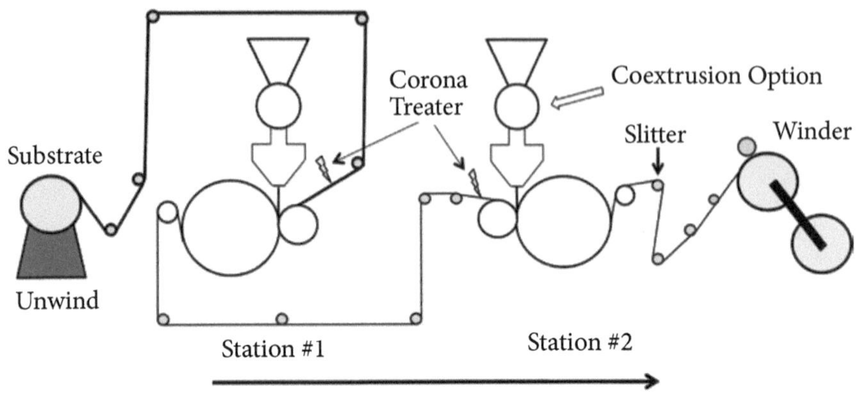

Figure 2.1

Two side coating tandem line

The split of the key extrusion coating application segments are shown in **Table 2.3**.

Table 2.3

Key extrusion coating markets

Liquid packaging	56%
Flexible packaging	17%
Industrial papers	14%
Other rigid	10%
Photographic	3%

Liquid packaging is the largest market sector for extrusion coating. The earliest large volume applications were in milk and juice cartons using 2-side LDPE coated paperboard. These were either gable top or brick shaped cartons. The tandem extrusion coating machine proved to be an ideal process for large-scale production at high output rates. The development of sterile aseptic packaging was the next major development. Paperboards coated with LDPE and incorporating an aluminium foil lining were found to have outstanding attributes for packaging fruit juices, UHT milk, yogurt, wine, soup and many other perishable edible products.

A few demanding commercial products are described below.

2.1 Paper/aluminium/polyethylene structures

One of the most versatile structures produced by the extrusion coating process at high-speed and low cost is the combination of paper, aluminium foil and polyethylene. The tandem extrusion coater is the most cost effective method of producing these products.

A typical machine is shown schematically in **Figure 2.2**. The substrate, a 40-75 g/m^2 lacquered and printed paper is first laminated to the aluminium foil with 10-12 g/m^2 of LDPE extruded at 315-325 °C. In the second station the heat seal layer of 20-40 g/m^2 LDPE is coated on the foil at 310-320 °C. Corona or flame treaters are required to ensure good adhesion between the polyethylene and the substrates. The resultant laminates are used on form, fill and seal packaging machines for liquid, powdered and granular products. Typical examples are soups, spices and cereals. Shampoos, fresheners and other liquid toiletries are also packed in sachets made from this type of laminate.

The addition of a coextrusion head at the second station enhances the versatility of the process by inserting thin layers of copolymers (EAA) with improved adhesion to aluminium foil. The schematic of the 2-layer coextrusion unit is shown in **Figure 2.3**. The satellite extruder (#2) is 60-70 mm in diameter and the main extruder is typically 150 mm (6″) in diameter. The EAA is essential for products that contain alcohols, vegetable oils and low acid foods. EAAs will develop excellent adhesion to aluminium at 285-295 °C thus substantially improving organoleptic properties. For hot fill applications the more costly zinc ionomer is sometimes recommended over EAA.

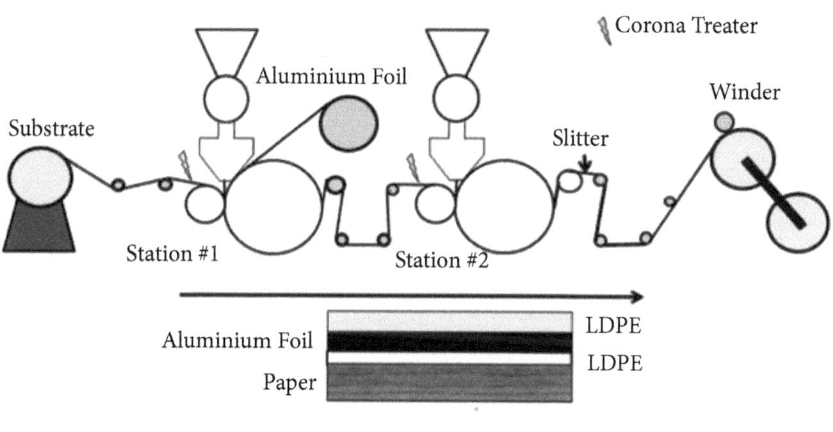

Figure 2.2

Tandem coating and lamination

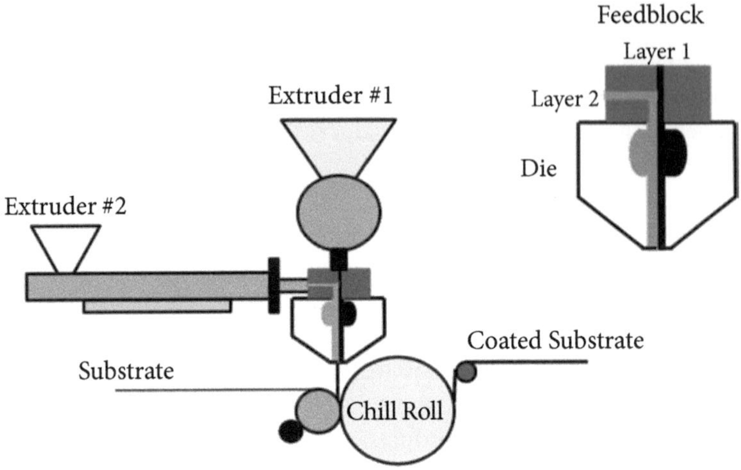

Figure 2.3

Two layer coextrusion coater

2.2 Triplex coating lines for aseptic packaging

The triplex (3 station) extrusion coating line was developed for aseptic liquid packaging of dairy and fruit juice products. The machine layout is shown in **Figure 2.4**. To enhance adhesion to the foil the inner LDPE layer is coextruded with an EAA copolymer as discussed above. A simple two layer coextrusion feedblock unit (see **Figure 2.3**) is placed at the third station where the adhesive resin is coextruded with the LDPE and coated on the foil.

The generic structure is:

Outside

LDPE
Paperboard
LDPE
Al foil
EAA
LDPE

Inside

A printed semi-bleached board is coated on the outside (printed side) with 10-12 g/m^2 LDPE in the first station. In the second station, 7-9 μm aluminium foil is laminated to the board with 10 g/m^2 LDPE, and in the third step the inside coating of 30-50 g/m^2 is made up of an EAA-LDPE coextrusion. The EAA ensures

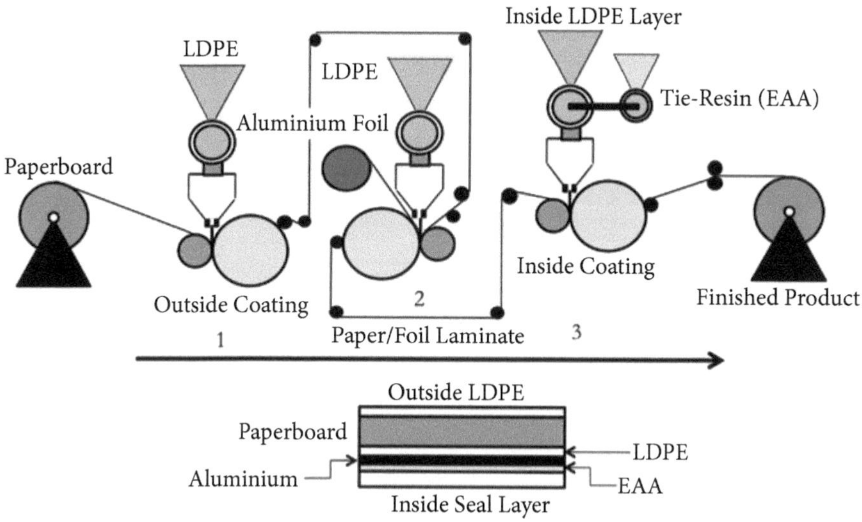

Figure 2.4

Triplex coating line for aseptic packaging

excellent adhesion to the aluminium foil and allows the use of lower temperature to minimize odour. To reduce costs the EAA coating is reduced to <10 g/m^2. This operation is carried out in a single pass at speeds reaching 400 mpm. Corona or flame treatment is required with all three steps. Flame treaters are preferred with large web widths of paperboard.

In some applications an inner layer with improved environmental stress crack resistance (ESCR) may be required. The typical 8 dg/min melt index extrusion coating grade LDPE will generally have poor ESCR. To overcome this deficiency tougher polyethylene films produced by the tubular film process can be extrusion laminated at the third station of the triplex line.

LLDPE has superior ESCR to LDPE and can also form stronger seals. The metallocene LLDPEs embody additional improvements in physical and heat seal properties plus the advantage of very low extractables (see chapter 13). Film coextrusions combining LD/LLDPE and mLLDPE make excellent lamination films in this context. The film is 35-45 μm in thickness and one side corona treated (40-45 mN/m) and extrusion laminated with 10-12 g/m^2 LDPE to the aluminium foil side of the structure.

It is important to minimize the presence of stabilizers, catalyst residues, waxes, slip agents and other impurities in the film to ensure the lowest odour and off-taste and compliance with food packaging regulations.

Typical commercial structures used for aseptic packaging of beverages are shown in **Figure 2.5** and **Table 2.4.**

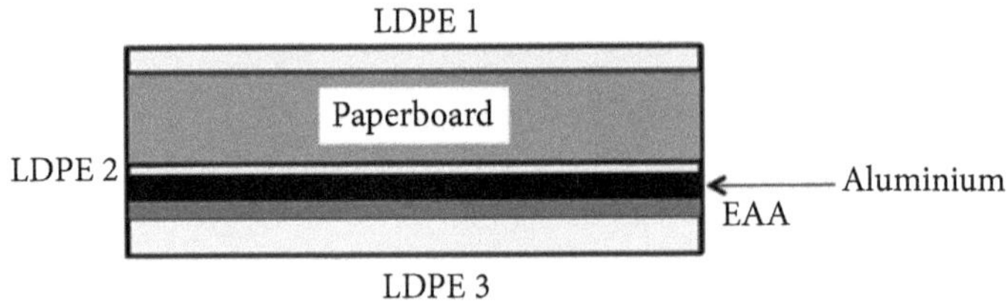

Figure 2.5

Aseptic liquid packaging structures

Table 2.4

Aseptic packaging structures

Material (g/m²)	Type 1	Type 2	Type 3
LDPE 1	25	14	17
Paperboard	245	300	260
LDPE 2	18	15	20
Al foil		7-9 microns	
EAA	9	7	9
LDPE 3	25	35	53

2.3 Photographic paper coating

Extrusion coating with polyethylene was successfully adapted to manufacture two side coated photographic paper. The PE coatings are applied principally to remove surface roughness and for waterproofing. Smoothness levels below 10 RMS (root mean square) are required for high quality images. Proprietary blends of LDPE and HDPE plus a titanium dioxide masterbatch are applied on the heavily sized and hydrophobic paper. A starch coating of 1-2 g/m² may be applied as a primer. Great care has to be taken with factory cleanliness and the elimination of gels and oxidized particles in the coatings: especially the side used for the photographic base. The LDPE coating weights are approximately 30 g/m². The coatings must be free from die streaks, chatter marks (intermittent irregularities) and pits. Temperatures of 315-320 °C are normally used at line speeds of 200 mpm or higher. HDPE (30-35%) blended with coating grade LDPE is applied on **one** side of the paper to counteract curling. HDPE grades of melt flow rates of 20 dg/min. at 230 °C and 0.962 g/cm³ density are used to provide the required increase in stiffness at the lowest blend ratio.

The development of digital photography has revolutionized the imaging industry by replacing the chemical processes with ink jet printing. The paper stock for ink jet printing is still essentially based on a two-side coated paper using LDPE and a blend of HDPE/LDPE as above. The main difference was to replace the gel subbing (adhesion promotion surface) layer on the face side of the paper with an ink jet receptive formulation. The quality specifications for the base photo-paper are believed to be less stringent than those used in conventional photographic processes.

Associated with the imaging industry is the requirement for thermal lamination film to protect and enhance products, such as photographs, prints and book covers. These films are produced by coating EVA (16-18% VA) on corona treated 12 μm OPET films. Special chill roll surfaces are required to eliminate blocking by the EVA (unwanted adhesion). In the heat lamination process the EVA melts easily and adheres to the substrate's surface to provide a tough and transparent protective layer with the OPET film (see section 1.2.)

2.4 Laminates for toothpaste tubes

Laminates are a low cost replacement for extruded aluminium tubes used for toothpaste and other viscous products. The advantages of laminates are lower cost and easier printability of the flat web. A typical structure is sketched in **Figure 2.6**. The adhesion requirements for toothpaste are stringent and require films that will not delaminate and stress crack. EAA copolymer is used as the laminating polymer because of its excellent adhesion to aluminium and polyethylene films. The outside film layer is pigmented white for enhanced appearance. The contact layer is also required to minimize the absorption (scalping) of the flavour substances such as mint and essential oils present in the paste. There is basically no solution to this other than minimizing the thickness of the inside PE film. Artificially spiking the paste formulation is the usual remedy.

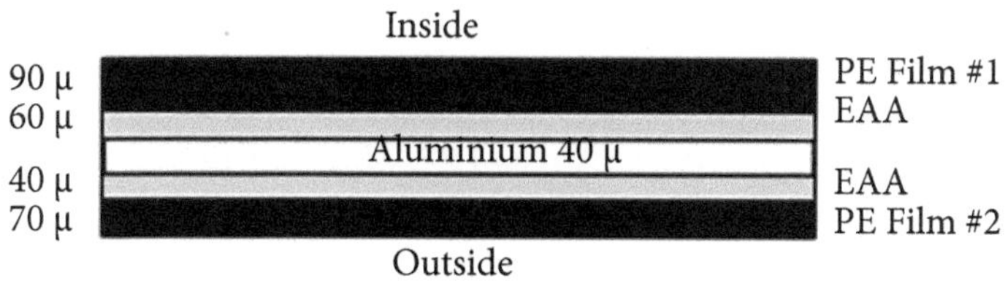

Figure 2.6

Laminate for toothpaste tubes

Figure 2.7 shows an installation used for the production of laminates for toothpaste and other viscous products. At station #1 a medium density polyethylene film is extrusion laminated with the EAA to the 40 μm aluminium foil. In the second station the outside layer of LDPE film is combined with the PE/EAA/Foil laminate to form a five layer final laminate. The EAA is crucial in ensuring good adhesion to both sides of the foil. An alternative solution is to coextrude the EAA with LDPE in the second station.

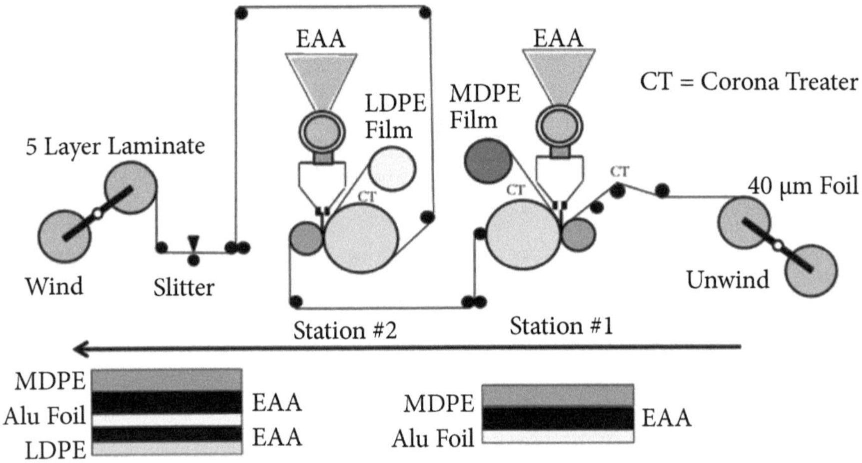

Figure 2.7

Extrusion coating layout for tube laminates

2.5 Barrier coatings and laminations

There are a number of patents, which describe the extrusion coating of EVOH (ethylene-vinyl alcohol) on paperboard (cardboard). The example in **Figure 2.8**[1] shows a combination that can replace aluminium foil and reduce scalping (loss of nutrients, aroma, etc.). Paperboard is extrusion coated on both sides with LDPE. The inner side is flame treated to raise surface tension. On a separate station a coextrusion comprising EVOH and LDPE combined with a tie-resin is extrusion coated on the flame treated LDPE side of the paperboard. The tie-resin is maleic anhydride (MAH) grafted PE, which works best with EVOH. The claim is that the molten EVOH will adhere to the flame treated LDPE layer. To produce this structure in one pass will require a triplex line (3 stations) with a 3-layer (ABC) coextrusion feedblock at the third station. The coextruded layer is typically 12 μm EVOH, 6 μm tie-layer and 18 μm LDPE. The presence of 18 μm LDPE will probably result in some scalping albeit reduced compared to a thicker coating.

1 Massouda D. F., U.S. Patent 5116649 **1992**.

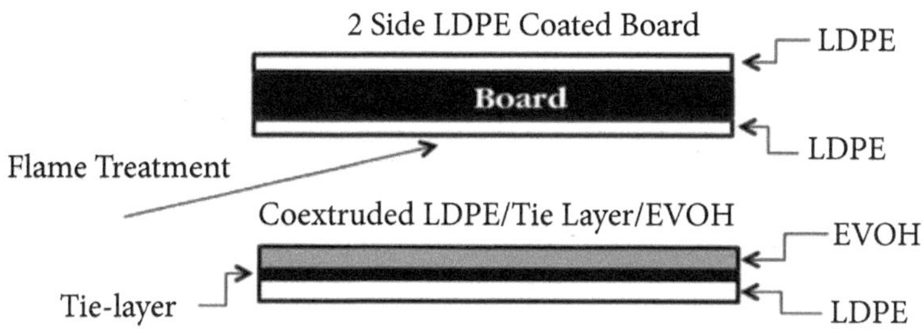

Figure 2.8

LDPE coated board with EVOH barrier

An addendum to this process is as follows. Treated polyethylene (surface tension: 45-50 mN/m) is wetted by aqueous and alcoholic solutions. EVOH is in essence an alcohol and should, therefore, interact with treated polyethylene. If this assumption is correct, a proposed alternative is to apply the EVOH as a monolayer on a flame or corona treated LDPE coated paper and to simultaneously extrusion laminate a treated PE film to the EVOH. This will eliminate the need for a 3-layer (ABC) coextrusion unit at the final station and the costly tie-resin is also eliminated. The assumption is that EVOH extruded at ~**280 °C** will bond to highly treated polyethylene. This solution can enable the production of paper-based structures that include an EVOH barrier and a polyethylene heat seal layer. The polyethylene lamination film can be blended from various combinations of LD/LLDPE, metallocenes etc. The film should preferably be free of any slip additives. The chill roll temperature could be raised to 30-40 °C, which should improve adhesion. A tandem line to coat the paper with an LDPE layer will be required at the first station. If this technique is viable the limiting factor will be the ability to draw down the EVOH melt without pinholing or edge break. The schematic layout is shown in **Figure 2.9**, which shows the second step of the process.

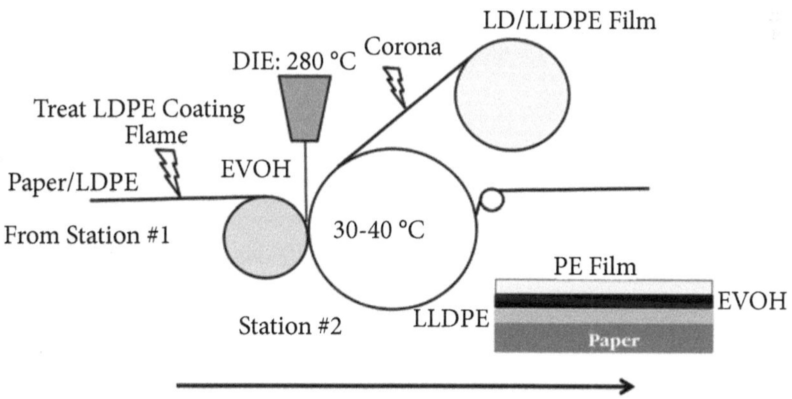

Figure 2.9

Proposed extrusion coating with EVOH

2.6 Extrusion coating PP on aluminium foil

Unless used with a very effective chemical primer polypropylene will only adhere to porous substrates and to itself. However, by coextrusion coating polypropylene (PP) with a **dual** function terpolymer good results are possible on aluminium foil and other smooth surfaced substrates. PP is coextruded with 5-10 g/m^2 of a terpolymer as the inner adhesive resin and coated on foil bearing substrates. The results in **Table 2.5** were obtained at 100 mpm with a 40 dg/min. melt flow rate PP homopolymer. The terpolymer was based on ethylene, acrylic acid and methyl acrylate. Air gaps of 110-150 mm were used. A warm chill roll of ~40 °C is recommended. Chill roll release was excellent from the PP layer.

Adhesion to the aluminium foil surfaces was good and no melt instabilities were reported. The bracketed figures are the adhesion values at the PP terpolymer interface. With some optimization of the terpolymer composition these experiments demonstrate that it is possible to extrusion coat PP on aluminium foil with good results. The presence of 6% acrylic acid in the terpolymer will also develop excellent adhesion to highly calendared and sized pulp and paper materials, which do *not* adhere to PP. The methyl acrylate level should be in the range of 6-9% b/w. Excess methyl acrylate will compromise the effectiveness of the acrylic acid. This coextrusion was easily run at 200 mpm.

Table 2.5

Adhesion of coextruded PP on Al foil

Substrate	Terpolymer (g/m^2)	PP (g/m^2)	Adhesion (N/15 mm @ 275 °C)	Adhesion (N/15 mm @ 285 °C)
Paper/foil	5	15	5.7	6.2
Paper/foil	5	35	4.6 (1.8)	5.9 (2.2)
Paper/foil	10	50	8.0 (2.0)	8.2 (1.8)
40 μm foil	5	35	-	NS (2.1)

A suggested application is to produce PP/terpolymer/foil/LDPE laminates in one pass on a tandem line fitted with a coextrusion coating head. A feedblock coextrusion head will deliver the terpolymer/PP layer onto the foil. The PP will embody high stiffness, very high gloss, grease and abrasion resistance and excellent release (non-stick) properties for a variety of pouch, sachet and liner applications. The LDPE can be replaced with EAA, which provides better adhesion at non-oxidizing conditions, and much stronger heat seals. This should allow for significant down-gauging of the inner layer. This product can replace BOPP/foil/PE laminates at a lower cost.

Paper-based structures typified below also offer potential:

PP
E-AA-MA
Paper
LDPE

These offer high gloss and a non-stick surface from the PP layer plus the benefit of the paper support.

Chapter 3

Extruder and screw design

In any flow channel a material will use as much of the channel area as it needs. If the melt is flowing too slowly as in **Figure 3.1**, part of the material will build-up and stagnate along the walls. This effect is normally referred to as channelling. Self 'scrubbing' from the walls will not occur if the applied shear stress is too low.

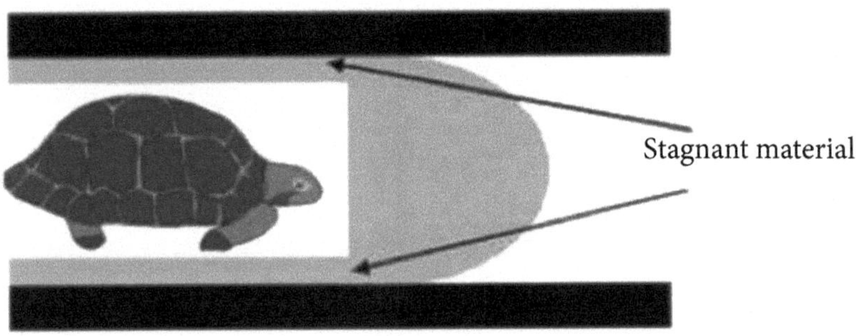

Figure 3.1

Stagnation of the melt at the walls

Machine designers will usually size pipes and other channels according to specified shear rates rather than calculating the shear stress, which actually determines the scrubbing action at the walls. The equations below show how to calculate shear rates for two simple flow fields.

Apparent shear rate at the lips of a flat die is calculated as in equation 3.1 where Q is the volumetric mass flow rate, W is the width and t is the die gap.

$$\gamma = \frac{6Q}{t^2(W+t)} \tag{3.1}$$

Apparent shear rate for a cylindrical section is calculated from the radius (R) as in equation 3.2:

$$\gamma = \frac{4Q}{\pi R^3} \tag{3.2}$$

Similar equations are available for annular dies. Some rules of thumb are used to estimate the minimum shear rates that will avoid channelling and polymer degradation along the walls. For LDPE the shear rate should not be below 20 s^{-1}. With heat sensitive polymers higher shear rates may be required to ensure that wall scrubbing is effective and residence time in the flow domain is minimized.

Figure 3.2 plots the shear stress developed in a pipe 400 mm in length. The effect of varying pipe diameter on shear stress is calculated for an LDPE and EVOH. The mass flow rate through the pipe is 500 kg/hr. Temperatures are 310 °C for the LDPE and 250 °C for the EVOH. As the diameter increases shear stress drops as expected. The EVOH flowing into the 10 mm diameter pipe develops a shear stress of 200 kPa, whereas the LDPE develops a much lower shear stress of 76 kPa in the same geometry. Note that as the pipe diameter increases the curves begin to merge.

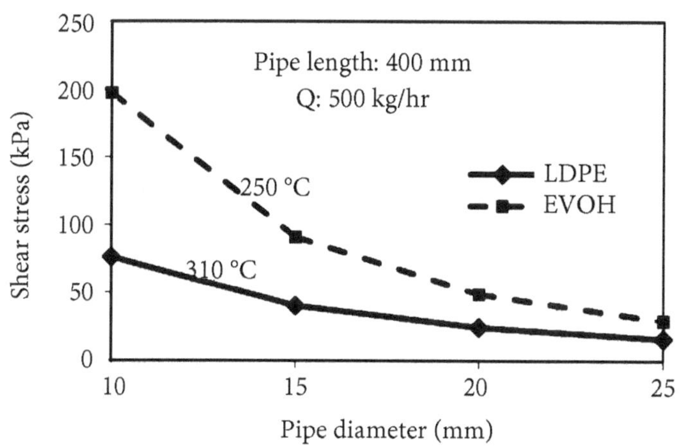

Figure 3.2

Effect of pipe diameter on shear stress

This simple example demonstrates the importance of the flow characteristics and processing temperatures of the materials to be used when configuring the geometry of all the flow paths. Excessive pressures and shear stresses must be avoided and conversely low shear/low velocity conditions will allow stagnation to occur at the walls of the flow channel. This becomes more critical with heat sensitive and potentially corrosive polymers. Costly polymers such as EVOH and adhesive tie-layers will usually be applied at very low thickness. Extruder outputs will therefore be **low** and flow fields should be designed to develop adequate shear stress to avoid wall stagnation. Problems like gels can form in regions such as the adaptor, connecting pipes, feedblocks and die edges where melt velocity may be low.

Sharp bends and flat entries into pipes and adapters must be avoided. **Figure 3.3** shows computer generated flow paths into two pipes with bends. The streamlines show stagnation at all the right angled bends. This is avoided by designing curved bends and tapered entries into the narrower channels as shown in the second pipe (no stagnation).

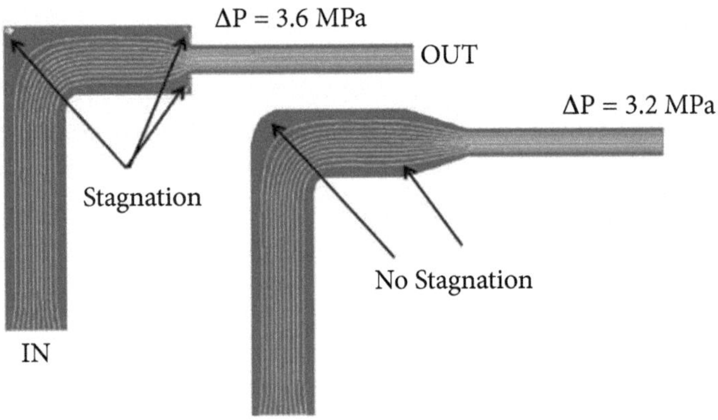

Figure 3.3

Melt stagnation at bends

These obvious principals may be overstated, but are often neglected when processors are confronted with malfunctioning machinery and tend to expend much effort diagnosing for more complex solutions.

3.1 Basic extruder functions

Figure 3.4 shows a typical extruder layout. The polymer is fed from the hopper into a heated barrel where the rotating screw melts and pumps the material through a screenpack filter and through the adapter into the die. In addition to filtering the screenpack converts rotational memory into longitudinal memory. Lever or ratchet operated automatic changers should be used to replace contaminated screenpacks without machine stoppage. In the die the melt at constant temperature and pressure is formed into the required shape. The extruder is fixed to a moving carriage, which drives the machine on rails to and from the coater. A hydraulic pumping device enables the lifting and lowering of the extruder over the chill roll. Sideways motion is also required to adjust the entry angle into the nip.

Heating is by electrical elements fitted to the barrel and independently controlled. Most extruders will have at least five separately controlled heating zones. In extrusion coating very high temperatures are used, particularly with LDPE, therefore, barrel cooling is normally unnecessary.

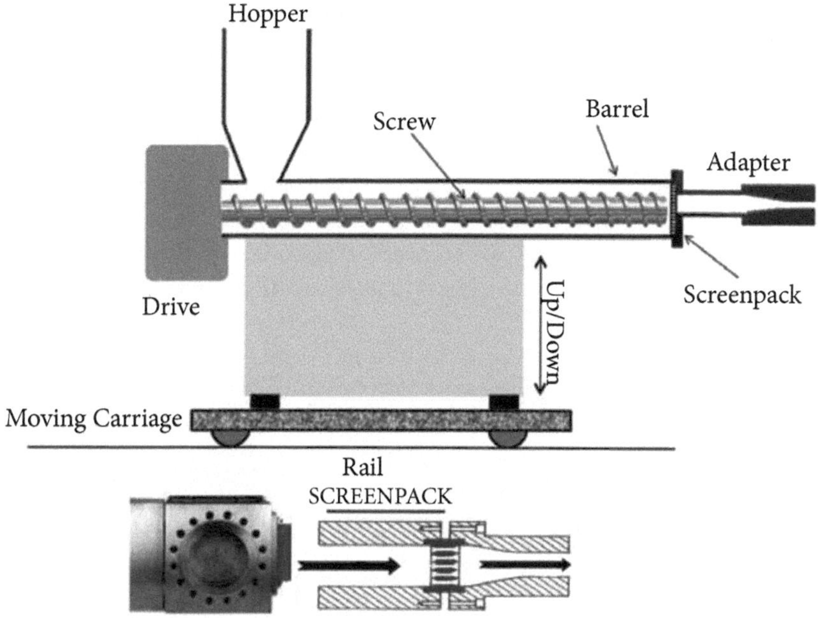

Figure 3.4

Basic extruder layout

The screenpack, made up of a breaker plate with tightly fitted screens, should be setup by placing the coarser screens upstream to catch large dirt particles. The finer screens are inserted progressively downstream.

Extruder capacity is matched to the die width, the coating weight and the line speed specified. The extruder must meet the following requirements:

- Minimum output variation (no pulsation)
- Constant melt temperature
- Minimum purging time during polymer changes
- Minimum residence time: avoid low screw speeds
- Minimum time for a step change from drool to the required operating output

The function of the coating extruder is to deliver the highest material output at a very high melt temperature. Since LDPEs of relatively high melt index

(4-8 dg/min.) with wide molecular weight distribution (MWD) are used, viscous dissipation is moderate. Extruder barrel temperature settings are high and the screw must be designed to develop a high degree of frictional energy. Fortunately, since the polymers used are highly shear sensitive, viscosity is low and high screw speeds can be used to maximize output without unduly overloading the drive.

An extruder will reach stable conditions from start-up faster if the mass of the barrel and screw are minimized. *The smaller the screw diameter the shorter the time constant for the machine to reach stable temperature.* The extruder diameter has a greater effect on the ability to reach stable conditions than its length, i.e., the length to diameter ratio (LD ratio). In other words avoid the use of unnecessarily large diameter extruders.

A well designed screw should be free of surging or output pulsations. High level pulsations can be picked up with pressure gauges and ammeters. Low level pulsations are more difficult to detect as pressure transducers may not measure real levels of melt flow fluctuations. The best method of detecting low level pulsations is to continuously monitor coating weight variations in-line. The following problems can all cause surging in extruders:

- Drive variations
- Gear slippage
- Worn screws
- Malfunctioning heaters
- Burned out fuses
- Corroded thermocouples

All the above will contribute to instability as well as poor temperature control, which is crucial for both thickness distribution and consistent adhesion to the substrate.

3.2 Origins of extruder output pulsations

Figure 3.5 shows the polymer flow in a single screw extruder. The solids are fed from the hopper into a conveying zone, which feeds the compression zone. In the compression zone the polymer is subjected to progressively increasing shear stress and begins to melt rapidly. The molten polymer is delivered to the metering section, which discharges it to the die via a connecting adaptor. The melt must be delivered at constant temperature and free of any surging or pulsations.

The key extruder design features that can influence output fluctuations are listed below:

- Feed hopper
- Solids transport (feed or conveying zone): A
- Melting zone (compression zone): B
- Melt conveying (metering zone): C

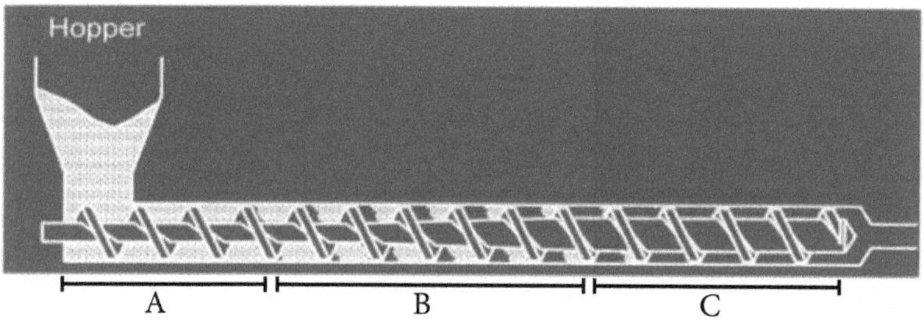

Figure 3.5

Melting process in an extruder

The process starts with the flow of pellets from the hopper followed by the melting and pumping action of the screw from the feed zone through to the compression and metering zones. **Figure 3.6** shows a simple hopper feed section. It is important that the slope and internal surface of the hopper allows the polymer pellets to flow freely and evenly into the throat and feed zone of the screw. Pellets should not funnel or arch at or around the throat resulting in uneven material flow. The

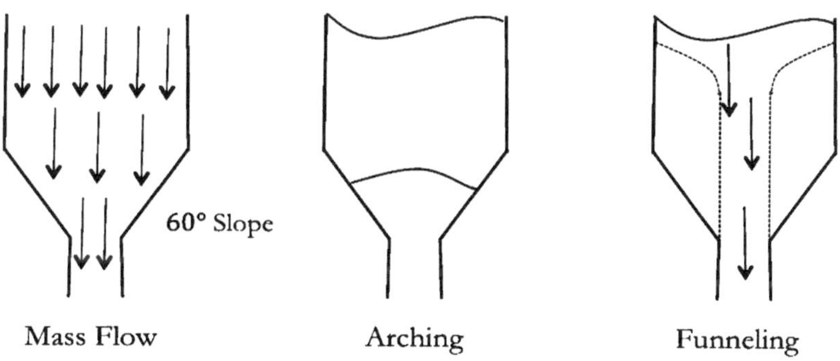

Figure 3.6

Hopper design

polymer silo storage and piping to the extruder hopper must be regularly checked for any build-up of fines, angel hair and dust, which may hinder the flow of pellets and cause cross contamination, resulting in gels in the melt.

In **Figure 3.7** the material must flow freely into the throat of the hopper and be carried away by the screw without melting, agglomerating or sticking to the screw. The pellets must be of uniform size and cut and dust free. Most extruders are made with a metal entry casting that includes the feed opening and the throat above it, as well as the surroundings for the first few flights of the screw. The casting must have water passages that cool all these areas. A temperature of 30-40 °C must be maintained in this zone. When extruding low melting polymers some designers recommend also cooling the screw root over the first few flights. Normally screw cooling is not recommended as it will reduce extruder output and waste energy.

The melting process is depicted in **Figure 3.8** (Tadmor model). The heat from the barrel and the friction between the barrel wall and the pellets initiate melting. A molten film is formed, which flows over the solid bed to form a melt pool at the rear of the flight. As the material flows downstream the melt pool grows until all the solids have disappeared. The completely melted material enters the final metering zone and is pumped into the die via the screen-pack and connecting adapter.

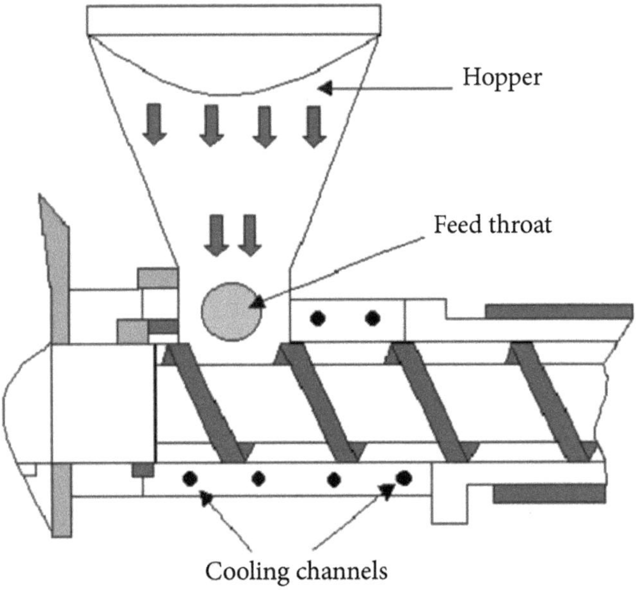

Figure 3.7

Cooling at the feed throat

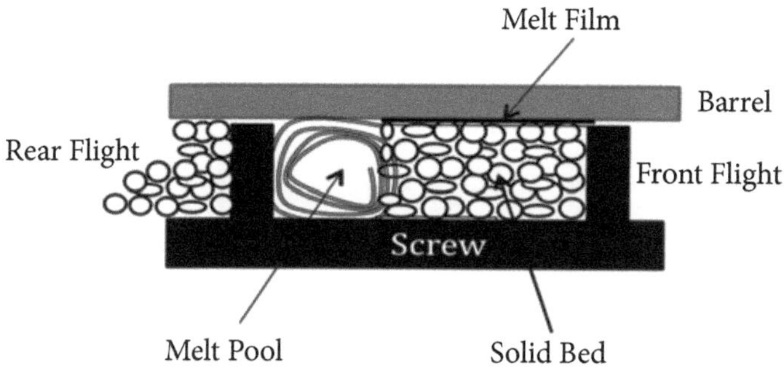

Figure 3.8

Melting in the compression zone (Tadmor model)

An important source of extruder surging originates in the **compression** zone. Solid bed wedging and solid bed break-up are two phenomena that can occur in this zone, which will lead to various degrees of surging. Both phenomena will usually originate from an inadequately designed compression section.

Figure 3.9 illustrates the formation of solid bed wedging. This occurs when the screw channel volume is reduced too rapidly. Not enough time is available to completely melt the material and part of the solids is wedged at the front of the flight leading to surging.

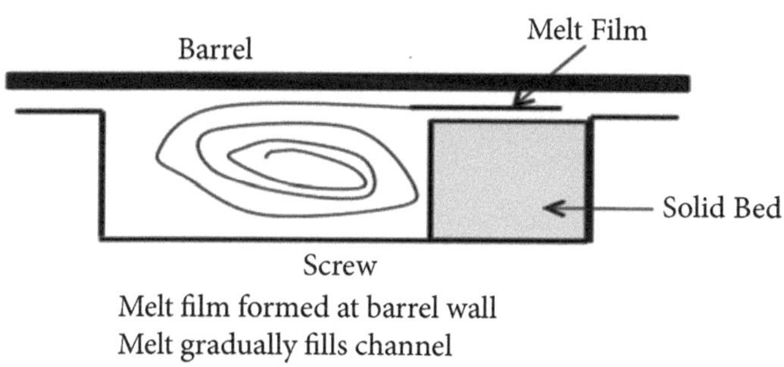

Figure 3.9

Solid bed wedging

In **Figure 3.10**, again due to a poorly designed compression zone unmelts may be formed as the solid bed breaks up. These unmelts swim in the melt pool, where shear stress is low, and will cause pulsation as they flow through the discharge end of the extruder.

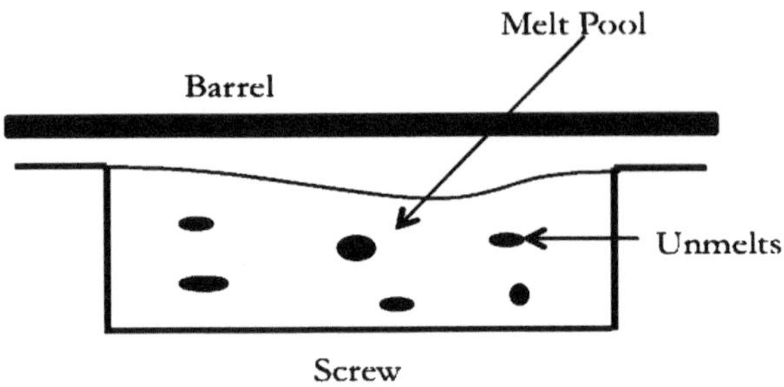

Figure 3.10

Solid bed break-up

3.3 Optimizing screw design

The key requirement of the extruder in extrusion coating is to deliver the homogenized LDPE melt at very high temperature. This is contrary to other processes, which are designed to operate at lower temperatures, e.g., the tubular film process. The economics, as always, specify the highest output and an LDPE melt temperature of 310-320 °C is normally required to generate sufficient adhesion.

The following aspects of screw design have been evaluated using computer modelling. The intent is to provide the reader with more insight into the critical features that govern good screw design:

- Melting (compression) zone
- Conveying (metering) zone
- LD (length to diameter) ratio
- Mixing section
- Extruder size
- Barrier screw
- Extruding polyester (PET) with LDPE screw
- Extruding EVOH with LDPE screw

The basic screw design features used in these simulations are as follows:

- Screw diameter: 114 mm
- Screw pitch: 1 D

- Flight width ~10% D
- Feed channel depth: 12.5 mm

The compression section must avoid the formation of solid bed wedging and break-up, which will lead to output pulsations as discussed above. The starting model is shown schematically in **Figure 3.11**. The extruder is 114 mm (4½") in diameter with a compression ratio of 3.6 : 1. The LD ratio is 28 : 1. The feed zone has a channel depth of 12.5 mm. The length of the compression zone is 6 D with the channel depth reducing to 3.5 mm. The metering zone of 15 D has a depth of 5 mm. The barrel temperature settings are shown in **Figure 3.11**.

The screw is run at 100 rpm with an 8.0 dg/min. melt index and 0.918 g/cm^3 density LDPE recommended for extrusion coating.

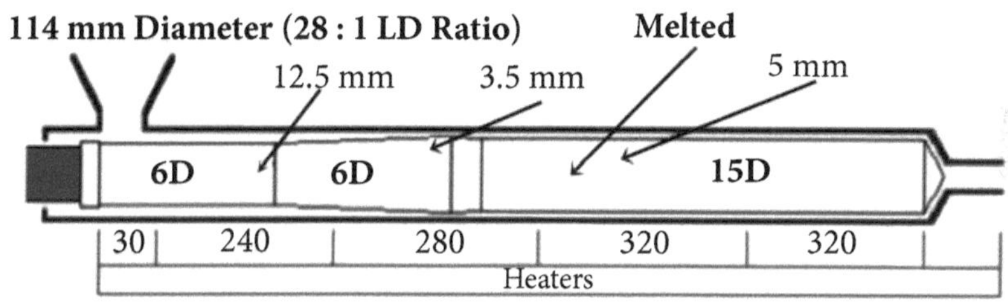

Figure 3.11

114 mm Diameter extruder

Figure 3.12 shows the plots for the temperature rise along the screw and the progression of melting by calculating the solid bed to melt pool ratio (**SB ratio**). The melting starts and begins to plateau as shown indicating poor melting. In extreme cases the curve will raise indicating solid bed expansion. This phenomenon is caused by insufficient volume and/or time in the compression zone. Eventually as the material flows downstream the solid bed ratio begins to contract roughly half way down the screw length. According to the simulation the material is fully melted at 16 D along the screw. The exit temperature is 293 °C and is too low for this application.

To improve the efficiency of the melting process the compression zone was expanded to 12 D. The new screw design is shown in **Figure 3.13**. The channel depths were unchanged.

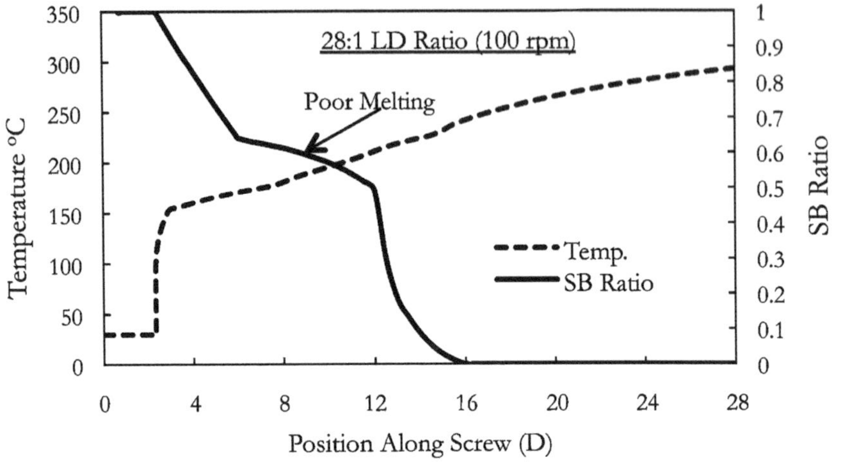

Figure 3.12

Solid bed expansion and temperature

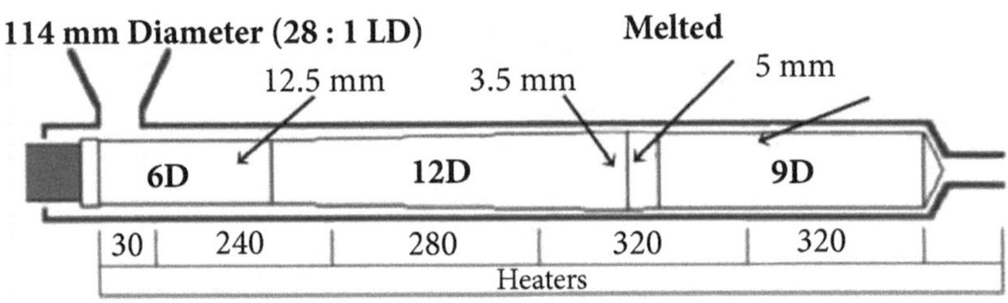

Figure 3.13

Extended compression zone

The outcome is shown in **Figure 3.14**. The solid bed is now consumed more efficiently in the compression zone. The risk of pulsation originating from this section of the screw is minimized. Temperature has dropped to <290 °C and is still much too low. This screw is **not** recommended for LDPE extrusion coating. The following sections discuss methods to increase melt temperature and maintain high output.

The next step was to generate a higher temperature by redesigning the metering zone. The aim was to increase viscous dissipation in this section of the screw. The modification is shown in **Figure 3.15**. The compression section was 8 D and the metering zone lengthened to 14 D. The total LD ratio was kept at 28 : 1 and the compression ratio was also unchanged. The screw speed was set at 150 rpm with the barrel temperature settings as shown.

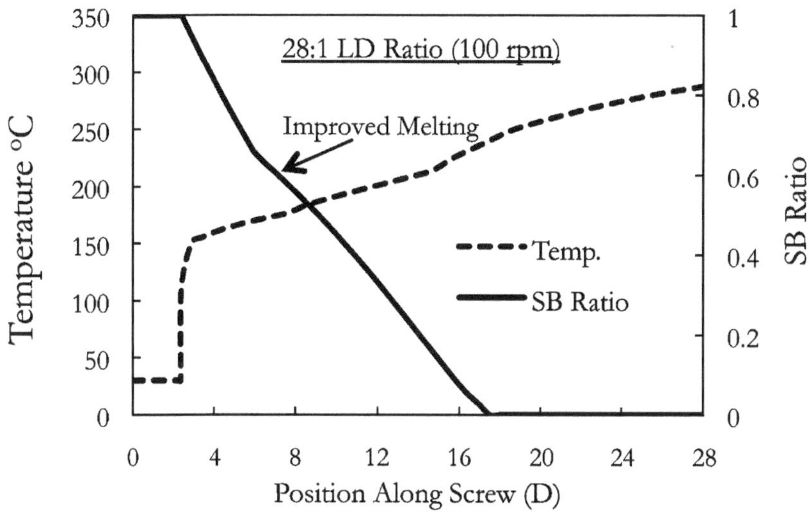

Figure 3.14

Effect of extended compression zone

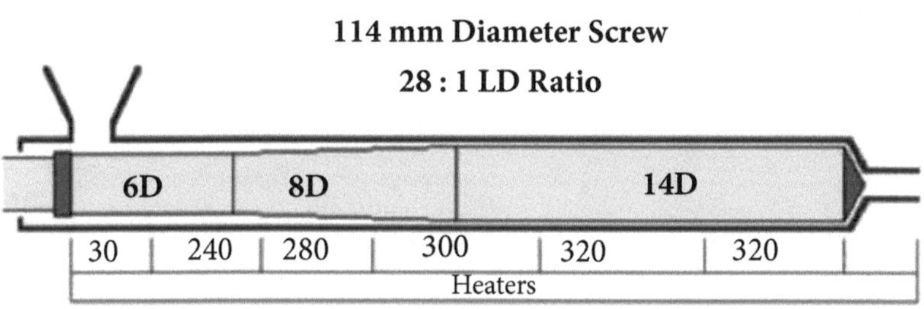

Figure 3.15

Extended metering zone screw

The channel depth in the metering zone was now varied from 2.5 to 4.5 mm. The results are plotted in **Figure 3.16**. As channel depth is increased output increases. Temperature, however, decreases. The graph shows that a metering channel depth of 3-3.5 mm is probably optimum for output and temperature. A compromise metering channel depth of 3.5 mm was, therefore, selected from which to explore further improvements in screw performance. With a metering zone depth of 3.5 mm the material is completely melted at 14-15 D. The interim screw design is shown in **Figure 3.17**.

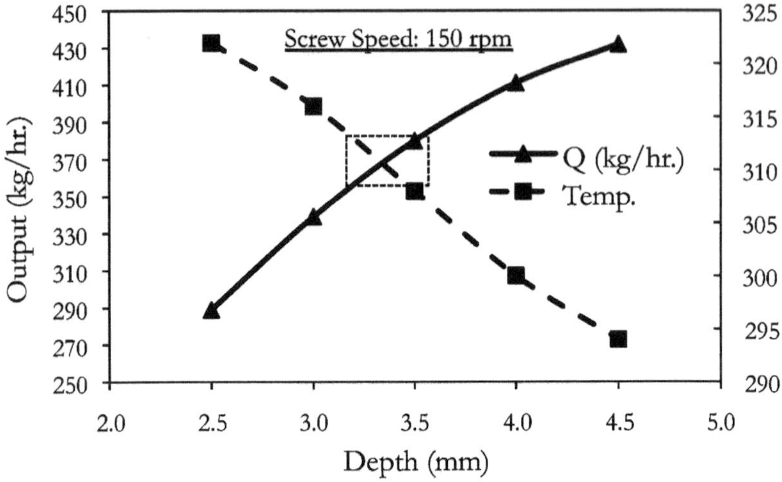

Figure 3.16

Effects of metering channel depth

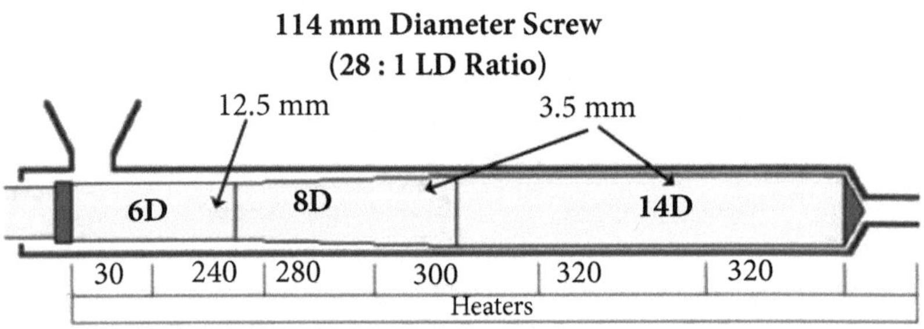

Figure 3.17

Simple screw for extrusion coating

Screw performance is shown in **Figure 3.18**. At 150 rpm this screw will deliver 409 kg/hr. of LDPE at a temperature of 307 °C. This average temperature is still on the low side of the recommended 310-315 °C. The solid bed ratio plot shows efficient melting. The peak pressure in the extruder is 12 MPa.

The data from these computerized studies indicate that this screw is incapable of reaching the desired temperature of 310-315 °C at high outputs with this LDPE. Reducing the channel depth in the metering zone will increase melt temperature, but the outcome is a loss in output per screw revolution. The simplest option considered was to increase the LD ratio from 28 to 32 : 1. The new longer screw profile is shown in **Figure 3.19**. Compression ratio is 3.6 : 1 as before. Channel depth at the feed zone was kept at 12.5 mm with the 15 D metering zone at 3.5 mm depth.

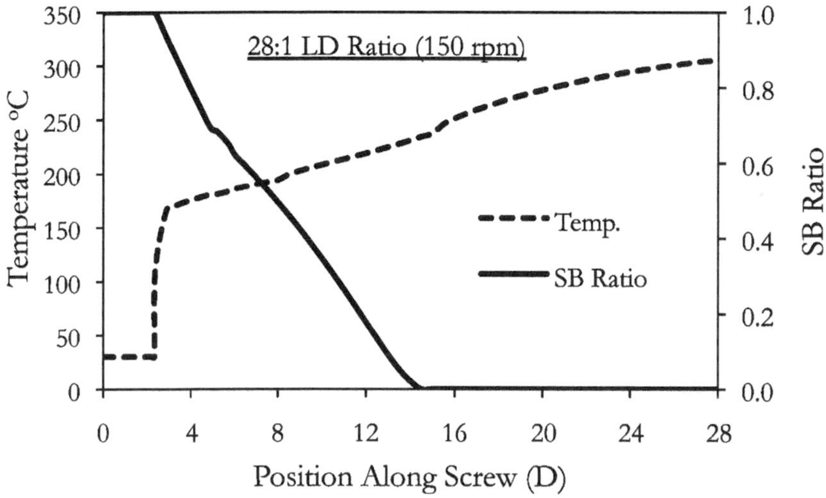

Figure 3.18

SB Ratio and temperature

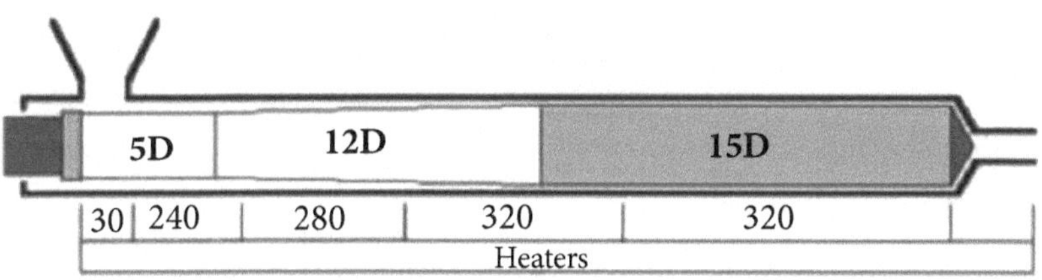

Figure 3.19

Profile of 32 : 1 LD ratio screw

The results with the same LDPE are summarized below:

- Screw speed: 150 rpm
- Output: 421 kg/hr.
- Average melt temp.: 311 °C

The average discharge temperature is now 311 °C. The higher LD ratio has enabled an increase in temperature with a small increase in output. The solid bed ratio and temperature profiles are plotted in **Figure 3.20**. The SB ratio plot shows the material melting smoothly in the compression zone with no signs that pulsations may originate from this section of the screw.

As a general conclusion LD ratios for extrusion coating screws for LDPE should be >**28 : 1** as demonstrated with this computerized evaluation.

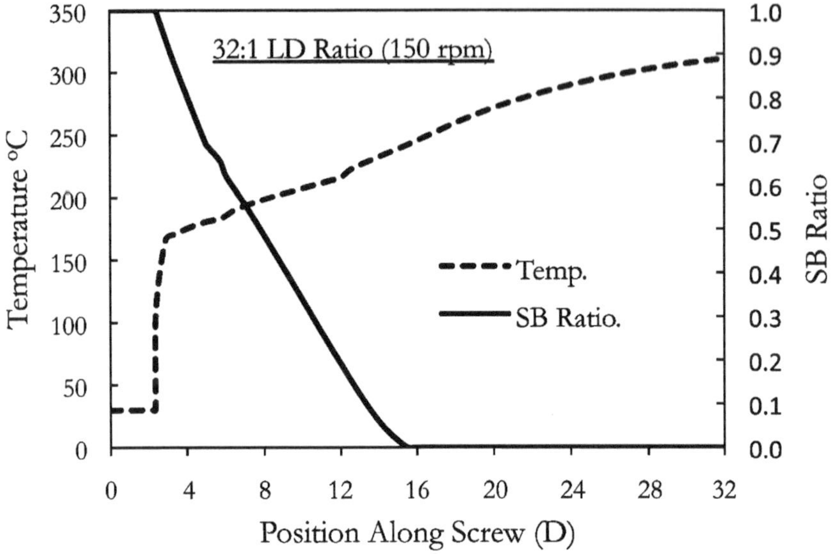

Figure 3.20

Temperature and SB ratio profiles

Single screw extruders have poor mixing capability. Mixing sections are, therefore, highly recommended particularly with polymer blends. Experiments with variable depth thermocouples have shown temperature variations ranging from 10-30 °C through the channel depth. The final modification was, therefore, to introduce a Maddock type spiral mixing section (2D) with a flight angle of 30° and a clearance of 0.5 mm. The section was placed 1D from the tip of the screw as shown in **Figure 3.21**.

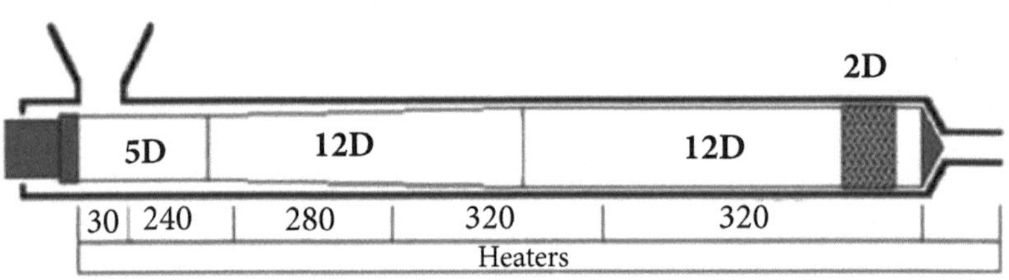

Figure 3.21

Profile of 32 : 1 LD ratio screw with mixer

In **Figure 3.22** the temperature distribution (ΔT) in a 3.5 mm deep channel is plotted at the inlet and exit of the mixer. The effect of the mixer is to smooth out the temperature resulting in a flatter curve except at the points nearest to the

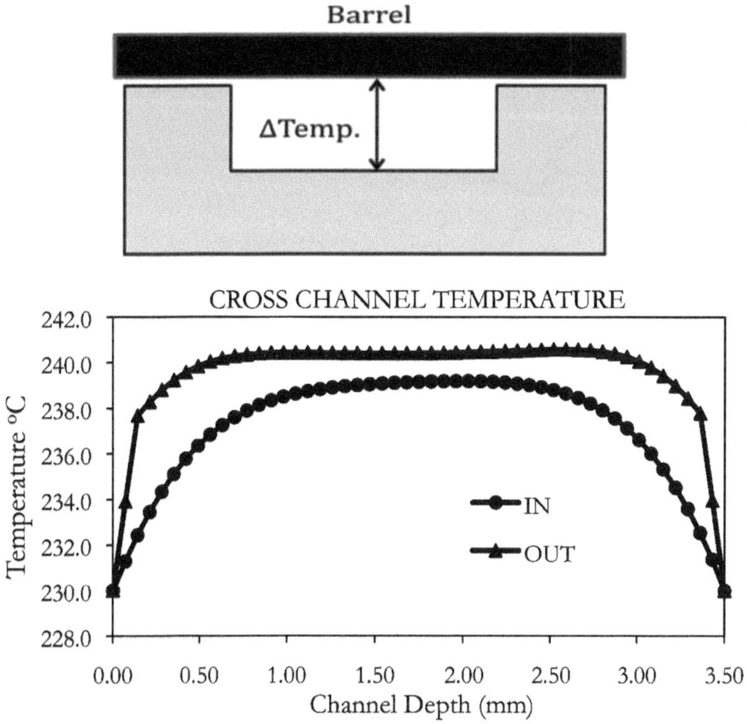

Figure 3.22

Effect of mixer on temperature distribution

barrel walls. In this example with the barrel wall setting of 230 °C the temperature rises to 239 °C in the middle of the 3.5 mm deep screw. With deeper flighted screws the variation can be much larger.

Many mixing designs are available: Maddock (fluted or spiral), pineapple, blister, pins, etc. Mixing sections are essential when using blends. Some designers will include two mixing sections.

Figure 3.23 plots the temperature and solid bed ratio profile for the screw with mixer from **Figure 3.21**. The melting process is efficient and the average exit temperature is 313 °C and within the target specification of 310-315 °C. The predicted output is 430 kg/hr. This stepwise approach has resulted in a screw of optimum performance for this application.

This extruder (**Figure 3.21**) was evaluated with a number of LDPE extrusion coating grades from different suppliers. **Table 3.1** tabulates the materials evaluated. The barrel temperatures were: 30/240/280/300/320/320 °C.

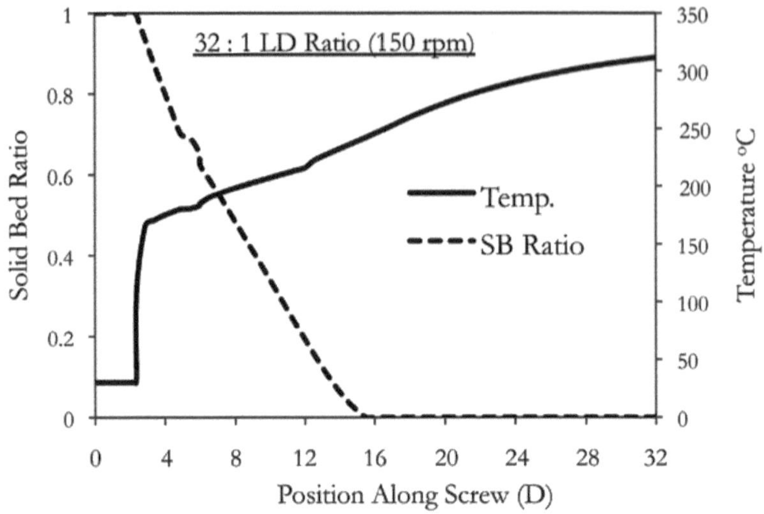

Figure 3.23

Temperature and SB profile with mixer. LD ratio: 32 : 1, compression ratio 3.6 : 1

Table 3.1
Commercial extrusion coating grades

Material	Melt index (dg/ min.)	Density (g/cm³)
LDPE (A)	3.5	0.923
LDPE (B)	8.0	0.918
LDPE (C)	7.5	0.918
LDPE (D)	5.5	0.917

Processability was assessed by determining outputs and average exit temperatures over a range of screw speeds. The results are in **Figure 3.24**. The predictions show that the screw's performance will vary with the choice of material.

LDPE (A) the 3.5 dg/min. melt index grade shows a rise in temperature with increasing output. The others, with higher melt indices, show the reverse. Resin (D) develops the lowest temperature. The curves tend to diverge as the output, i.e., screw speed, increases. **This is important**: as an example at 500 kg/hr. resin (C) develops a temperature of 307 °C whereas (B) reaches 312 °C. This may result in poorer adhesion and lower levels of smoke and odour with (C). LDPE (A) reaches a temperature of 325 °C and should develop higher adhesion than (B) and (C). These observations illustrate the need for good rheological data to be able to predict the performance of each material prior to final selection. Temperature

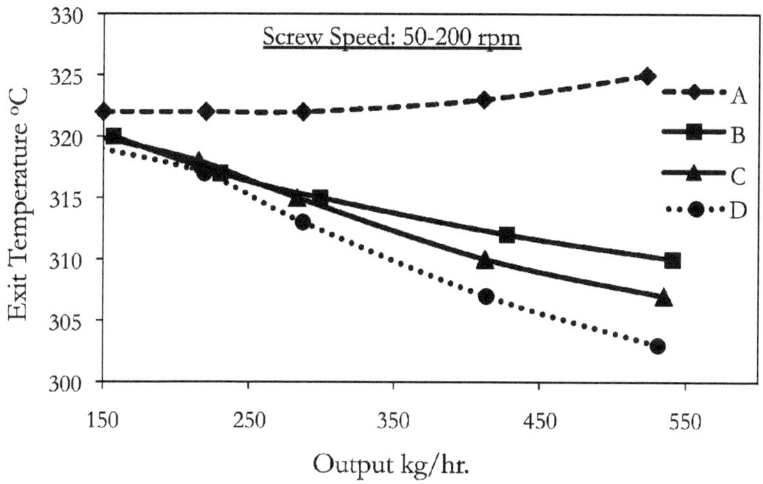

Figure 3.24

Comparative processability

measuring devices are not always reliable in detecting small temperature changes particularly on large machines.

Note*. The rheology data for the materials entered in Table 3.1 were obtained from different sources. Although the Carreau curve fitting viscosity model was used in all four examples different measuring methodologies are used in different laboratories, which may reflect on the calculated results given here.*

3.4 Barrier screws

Smooth bore **barrier** screws are widely used to extrude a variety of different polymers. These screws include a dual flighted section, which separates the solids from the melt. The intent is to improve the efficiency of the solids conveying and melting process thus increasing output per revolution. The principle of the dual flighted barrier screw is shown in **Figure 3.25**. The solids are separated in one channel where melting commences from heat and friction at the barrel wall. The molten material flows over the secondary flight, designed with a specified clearance, and accumulates in the secondary channel to form the melt pool. The volume of the secondary channel is progressively expanded and the primary (solids) channel reduced as the melting process develops downstream. For maximum efficiency the primary channel must remain full of solids as its volume contracts in the downstream direction.

A correctly designed barrier screw should ensure that the primary channel is always full with solids. Two basic designs are used. The Hartig Barr design is to progressively reduce the depth of the primary channel held at constant width while the melt channel depth increases. This offers the maximum surface for the shearing action of the solids on the barrel to occur. The alternative is based on the Maillefer travelling barrier design, which continuously reduces the solids channel width at a constant depth. Combinations of the two concepts have also been designed.

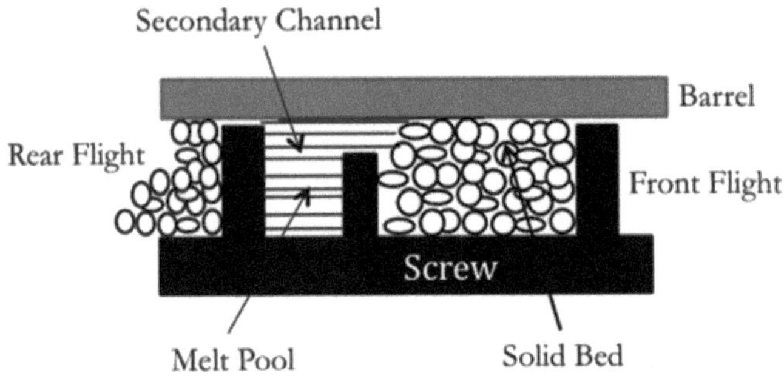

Figure 3.25

Dual flighted barrier screw

The next set of evaluations were carried out by modifying the basic 114 mm diameter screw (28 : 1 LD) by replacing the compression zone with a barrier section 11 D in length (see **Figure 3.26**). Dual flighted entry and exit sections of 1 D each were included. The clearance for the secondary flight was 1.0 mm. The screw pitch was kept at 1 D, i.e., square pitch along the whole screw. The melt channel (secondary) width in the barrier section was 20 mm and its depth is increased from 1 to 5 mm along its length. The feed and metering zones and fluted mixer were unchanged. This is by no means an optimized barrier design and is used simply to illustrate alternative possibilities. The results with a coating grade LDPE are summarized in **Table 3.2**.

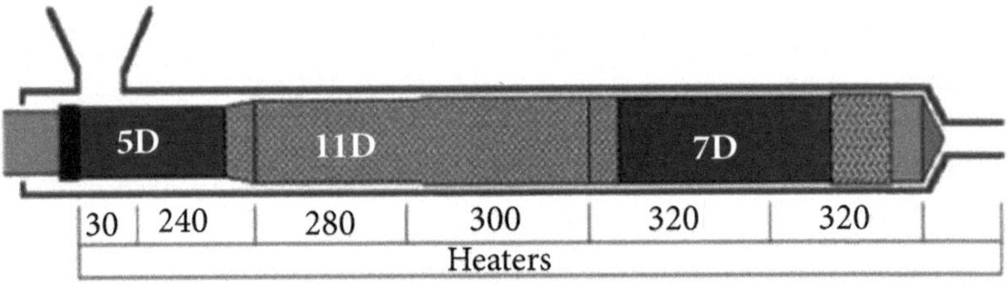

Figure 3.26

114 mm diameter screw with barrier section

Table 3.2

Results obtained for a 114 mm diameter barrier screw, LD ratio 28 : 1

Results @ 150 rpm	
Output (kg/hr.)	409
Exit temp. (°C)	311
Peak pressure (MPa)	8.4
Fully melted (D)	12

The output is 409 kg/hr. and the exit temperature is 311 °C. The solid bed ratio along the length of the barrier screw is plotted in **Figure 3.27**. The feed section is full with solids, which then drops in the transition flight to the barrier section. As the primary channel fills up with solids rapid melting occurs and the melt flows into the expanding secondary channel, which leads into the metering zone. Melting is very rapid in the primary (solids) channel. The material is fully melted at 12 D as it flows into the metering zone starting at 18 D. The simulation indicates that extrusion coating grade LDPEs extruded at very high temperature will probably not benefit much from the use of the more costly barrier screw. Melting is very rapid at these very high temperatures unlike other processes, which operate at much **lower** temperatures using low melt index polyethylenes.

Barrier screws are now standard in extruders used to produce polyethylene films by the tubular (blown) film process.

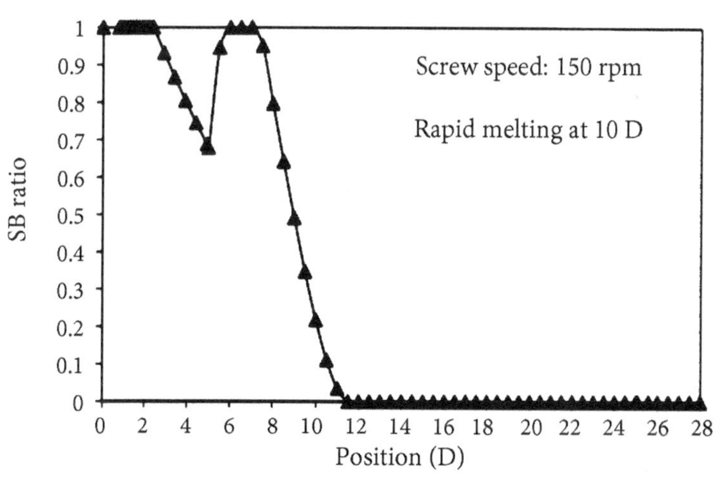

Figure 3.27

LDPE Melting in barrier screw

3.5 Sizing extruders

The size (diameter) of the extruder will affect pressure, temperature and residence time at any given set of process conditions. Shear rates over flights will increase with screw diameter and may cause localized overheating. These parameters will have some effect on adhesion, odour and gel formation.

The 114 mm, 28 : 1 LD ratio screw was compared to a similar screw with a diameter of 152 mm. The outline of the larger extruder is shown in **Figure 3.28**. The required screw speeds and estimated residence times at an output of 450 kg/hr. were computed for both extruders using the same grade of LDPE. The results in **Table 3.3** show the larger extruder with a longer residence time for the same output. Some air is always entrapped with the pellets entering the extruder, which will initiate some polymer oxidation. The longer the residence time the more likely polymer degradation becomes.

Table 3.3

Influence of screw diameter on required screw speed and residence time (for 450 kg/hr. output)

114 mm Extruder	152 mm Extruder
Screw speed: 167 rpm	Screw speed: 84 rpm
Residence time: 143 s	Residence time: 288 s

Figures 3.29 and **3.30** show the pressure and temperature profiles developed in both extruders. A higher pressure peak and temperature develop in the larger machine. The average exit temperature differs by 8 °C. This could affect adhesion and odour all else being equal. More heat sensitive polymers may be more prone to degradation in the larger machines as a result of higher temperature, higher shear rate and longer residence time. Gel formation has been reported with ionomers processed in large extruders.

The study demonstrates the care that must be taken when attempting to replicate a set of process conditions on extruders of different size and provenance. The

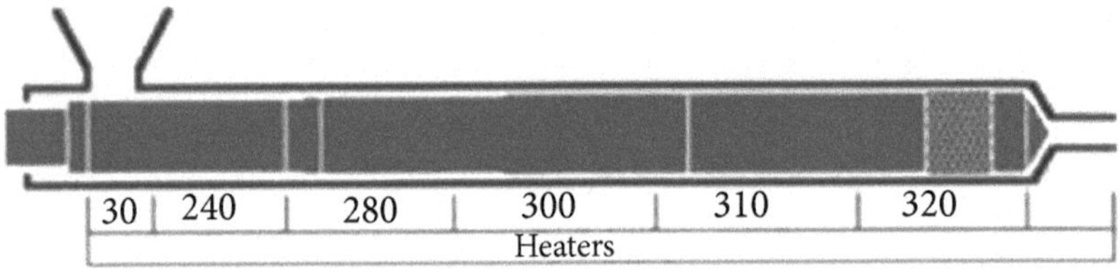

Figure 3.28

152 mm diameter screw

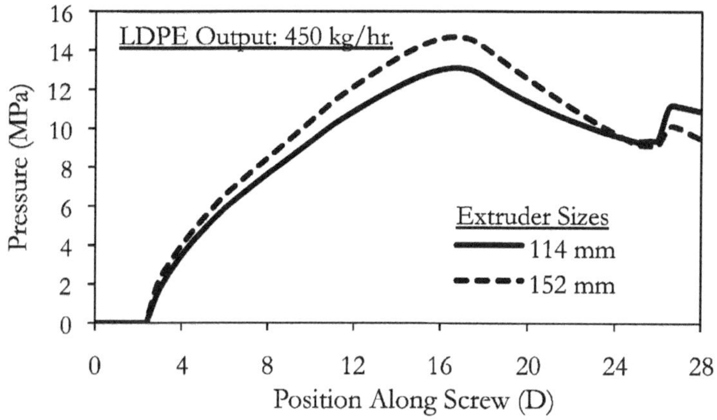

Figure 3.29

Effect of extruder size on pressure

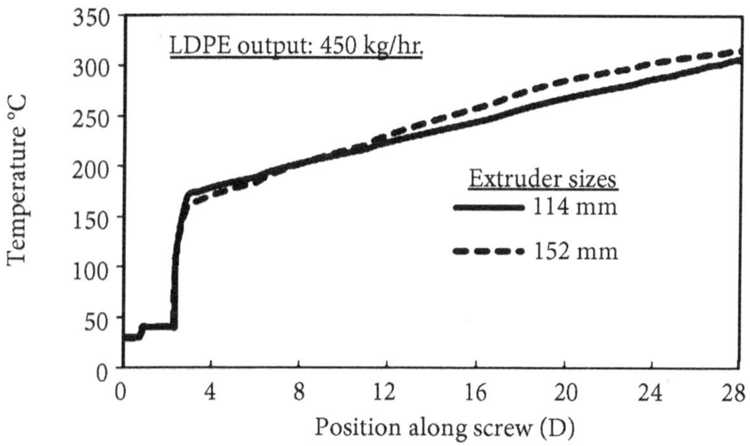

Figure 3.30

Effect of extruder size on temperature

assumption tends to be that if all the processing parameters are replicated with the given material then the outcomes will be identical. This presumption can lead to unexpected problems.

Table 3.4 lists the performance of a few commercial extruders with a range of polymers. Although extruders of 200 mm (8") diameter and higher are offered these large machines may lack operational flexibility. Extruder sizes of 90-114 mm (3½-4½") will take up to 20-40 minutes to reach equilibrium. Larger machines may take up to two hours.

Degradation problems have been reported with ionomers and other copolymers processed at high temperature in large extruders. If outputs greater than that specified from a 150 mm (6") diameter extruder are required it is suggested to consider the alternative of boosting output with a satellite extruder in conjunction with a coextrusion feedblock arrangement. Coextrusion will offer greater flexibility in controlling output than the installation of very large extruders.

Table 3.4

Extruder outputs (kg/hr.) for various film polymers

Extruder size (inches/mm)	Drive (kW)	LDPE	LLDPE	HDPE	PP	Ionomer	EAA	EVOH	PET
1½ (38 mm)	15	50	35	45	35	30	35	20	40
2 (50 mm)	40	115	70	90	70	60	70	45	80
2½ (60 mm)	60	180	110	145	110	90	110	70	125
3½ (90 mm)	120	350	210	280	210	175	210	135	250
4½ (114 mm)	200	650	390	520	390	325	390	250	450
6 (150 mm)	350	1150	690	920	690	450	-	-	800

3.6 Polyester (PET) extrusion

High viscosity copolyesters are available for extrusion coating applications (see section 14.3). These are dried in-line with dry hot air at a temperature of 60-65 °C to guarantee moisture content below 100 ppm. For extrusion coating the recommended processing temperatures for copolyesters are 280-290 °C. At these conditions the coating is amorphous and will be clear and heat sealable. Neck-in will be around double that experienced with LDPE. Fiber tear adhesion to paper substrates at coating weights down to 25 g/m^2 and a line speed of 150 mpm is claimed by the manufacturers. On non-porous substrates such as aluminium foil a chemical primer or coextrusion with an adhesive resin is necessary.

Figure 3.31 compares the shear viscosity of the copolyester to the LDPE at the recommended extrusion temperatures. The differences in shear viscosity are very substantial as shown. Significantly the slope of the polyester curve is less pronounced at the high shear rate part of the curve. Melting points are >200 °C compared to ~110 °C for LDPE.

A commercial 114 mm diameter screw from a well known machinery constructor was replicated in the computer software and run with this grade of copolyester. The screw is shown in **Figure 3.32** and includes a fluted mixing section. The copolyester (intrinsic viscosity: 0.75 dℓ/g) was preheated to 65 °C and fed to the extruder. The barrel temperatures are shown in **Figure 3.32**.

At 100 rpm the extruder delivers 400 kg/hr. of copolyester. The temperature and pressure profiles are shown in **Figure 3.33**. The polymer exits the extruder at 300 °C and the peak pressure of 30 MPa is more than double that generated by LDPE. This high pressure may lead to leaks at flanges if the machine is not well maintained.

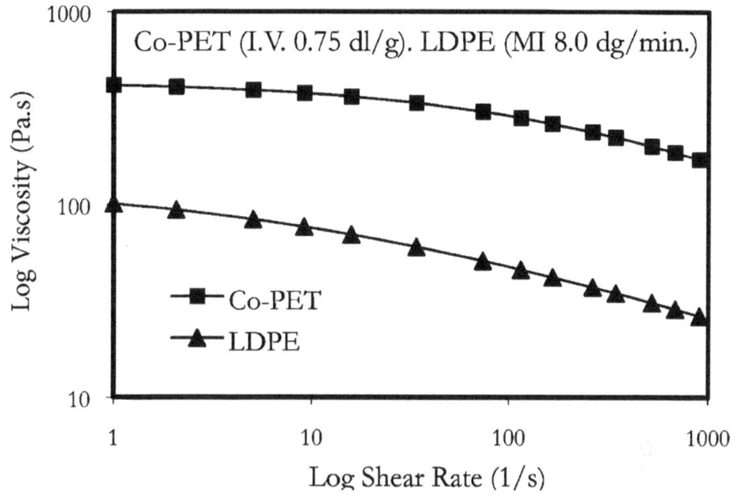

Figure 3.31

Shear viscosity of CoPET (290 °C) and LDPE (310 °C)

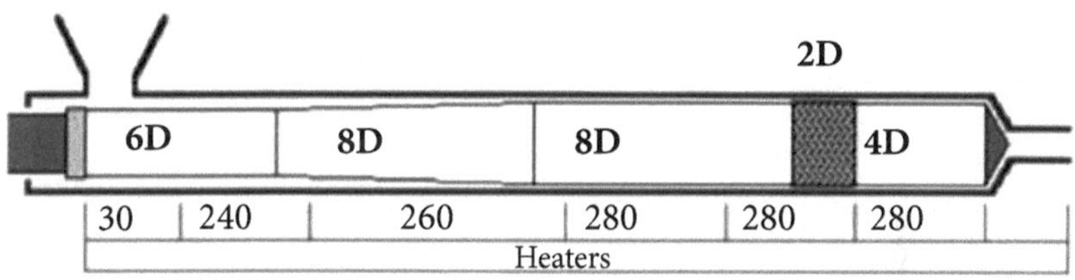

Figure 3.32

114 mm diameter 28 : 1 LD ratio extruder, compression ratio 3.6 : 1

The temperature rise at the exit is 20 °C higher than the set point in the metering zone. **Figure 3.34** plots the solid bed profile along the 28 : 1 LD ratio screw. The polymer is fully melted at 17 D. The melting appears to be fairly smooth although the high melting point of this resin compared to LDPE requires more time in the compression zone. This is indicated by the shoulder of the SB ratio curve.

The manufacturer claims that this specific copolyester will have double the neck-in of LDPE at the same conditions. The claim is that this grade can be run at line speeds up to 150 mpm without instability or draw resonance.

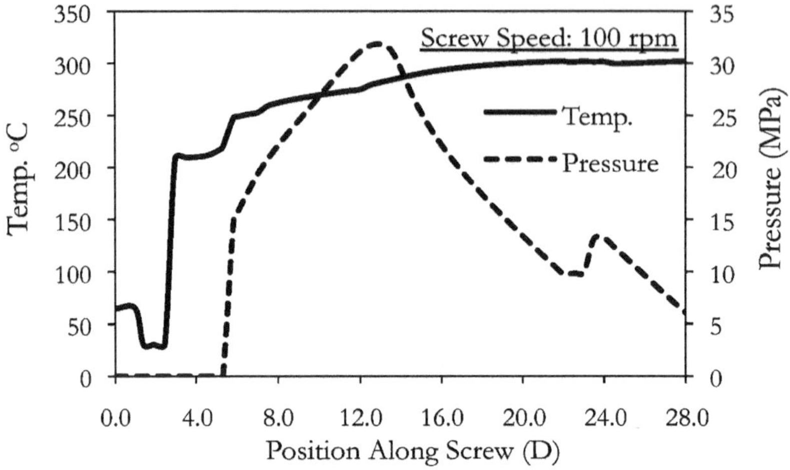

Figure 3.33

Copolyester temperature and pressure profiles

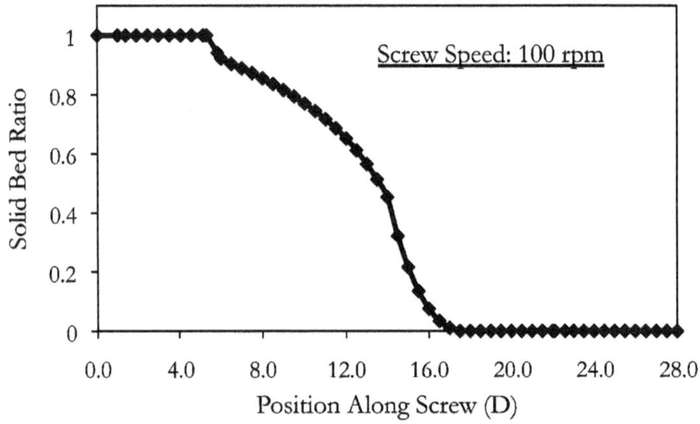

Figure 3.34

Copolyester solid bed profile

3.7 Ethylene-vinyl alcohol extrusion

EVOH will normally be coextruded between polyolefins to provide moisture protection. EVOH resins are heat sensitive and should not be extruded at temperatures >250 °C. All flow channels must be designed to generate a high enough velocity to eliminate stagnation in the flow domains. More recently special grades for extrusion coating have been developed. These are claimed to be processable at temperatures up to 280 °C.

EVOH layers are usually <10 µm in thickness and processed in relatively small satellite extruders on a coextrusion machine. A small 50 mm diameter extruder used by a well known EVOH supplier is replicated in **Figure 3.35**. The screw has a feed zone of 6 D and channel depth is 9.9 mm. The 10 D compression zone has a channel depth increasing from 2.76 to 9.9 mm. The 10 D metering section is 2.76 mm deep. The LD ratio is 26 : 1 and compression ratio is 3.6 : 1.

Table 3.5 shows the outputs at three screw speeds. These compare the actual experimental results from the supplier and the PC simulated data using the design information from **Figure 3.35**. The computer generated output data show an excellent match with the actual measured outputs.

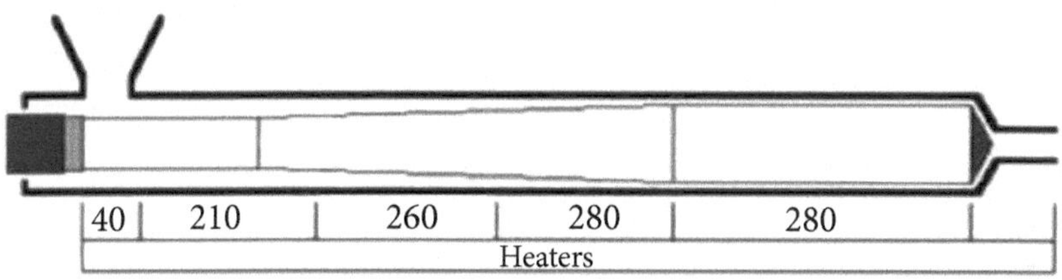

Figure 3.35

50 mm Diameter extruder for EVOH, compression ratio 3.6 : 1

Table 3.5

Screw performance with EVOH

Screw speed (rpm)	Actual output (kg/hr.)	PC simulated output (kg/hr.)
11.5	7.1	6.3
48.9	26.9	27.6
97.8	52.3	52.0

The same EVOH was run on the 114 mm (LD ratio 28 : 1) diameter commercial screw shown in **Figure 3.32**. This is a typical extrusion coating screw for LDPE. The barrel temperature settings were 30/210/260/280/280 °C.

The temperature and pressure profiles at 100 rpm screw speed are in **Figure 3.36**. A pressure peak of 20 MPa is developed and the average exit temperature is 287 °C. Output was 325 kg/hr. at 100 rpm screw speed.

The solid bed profile for the above screw and the 50 mm diameter screw are compared in **Figure 3.37**. The larger extruder exhibits quicker melting in the compression zone.

This basic commercial extruder performs well with this grade of EVOH polymer, however, as with the copolyester the EVOH requires more time in the compression zone to completely melt before reaching the metering zone in spite of the 280 °C extrusion temperature used. At 100 rpm the temperature rises some 7 °C above the set point.

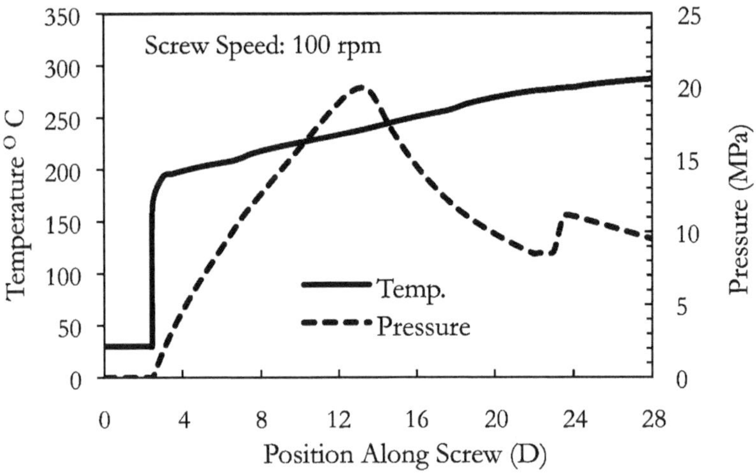

Figure 3.36

EVOH Temperature and pressure profiles (114 mm diameter screw)

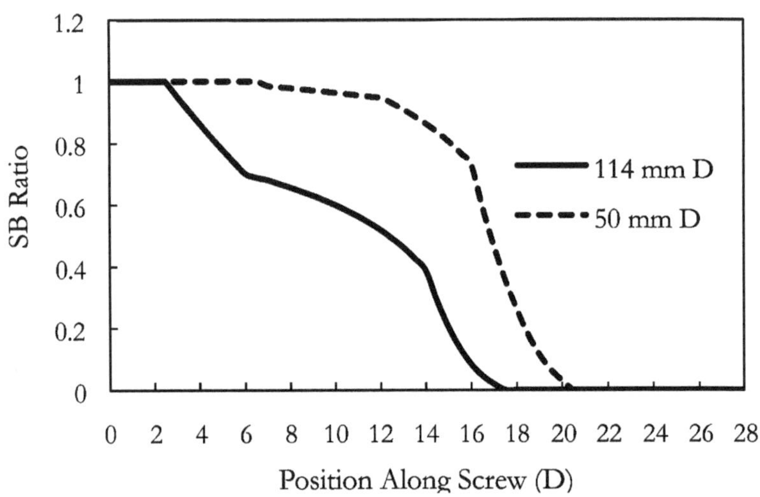

Figure 3.37

EVOH Solid bed profiles for 2 different diameter screws

Chapter 4

Die design

The underlying principle in die design is to avoid cross flows, which lead to uneven flow rates across the exit channel. The melt will take the shortest path from entry to exit and will flow preferentially down the middle. The key to good die design is to balance the pressure across its width to enable an even distribution of melt from the centre to the edges. **Figure 4.1** shows a typical coat hanger die with material flowing from the inlet port into the feed channel through to the restrictor (manifold) from which it should be evenly spread across its width. The final section is the die lips, which form the melt into its final shape.

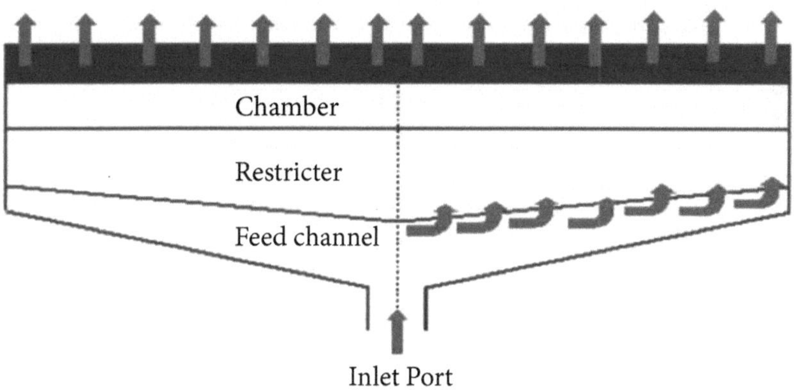

Figure 4.1

Balanced melt flow inside die

A correctly designed die should meet the following basic criteria:

- Balance the flow for uniform thickness distribution
- Maintain temperature homogeneity
- Maintain a reasonable pressure drop
- No hot spots
- No stagnant flow
- Avoid die lip build-up

- Avoid sharkskin and melt fracture
- Minimize residence time (maximize flow velocity)

Extrusion coaters use wide dies and line speeds which may be in excess of 300 mpm. Since substantial volumes of polymer will pass through the die per unit time, large extruders are used and die widths of 2.5 m and above are not uncommon. Extruder outputs for paper/board coating machines will range from 500-1000 kg/hr. **Table 4.1** lists typical die sizes with the best matching extruder.

The coating die opens to the outside through a straight line shaped orifice through which the melt is forced at constant flow. Good die design should be such that the polymer flow is streamlined with no chance of even tiny particles being trapped and overheated and degraded.

Table 4.1

Extruder sizes and widths

Extruder size	Die width
90 mm (3½")	600 to 1200 mm
114 mm (4½")	900 to 2000 mm
150 mm (6")	1200 to 3500 mm
200 mm (8")	up to 5000 mm

The extruder and die are set on a carriage, fixed on rails, which allows movement to and from the coater. Melt thickness is rapidly reduced as the melt is drawn and accelerates from the die into the coater where it is quenched. It is important to be able to accurately control the distance from the die lips to the nip and the angle at which the melt enters the nip.

Figure 4.2 shows the schematic cross section of two basic types of flat die concepts. These are defined as **T slot** and **coat hanger** die respectively. The T type is characterized by a constant feed channel cross section and the coat hanger design by a decreasing cross section. The latter's distributor or restrictor will decrease in length from the centre to the edges to around half to one third of its original surface area to form the coat hanger, i.e., distributor profile. The end plates as shown in **Figure 4.3** are bolted to the edges of the die. Any leakage and hang-up must be avoided. The end plates must be precision machined and the distributor insert must be designed to ensure streamlined flow and avoid any hang-up at the edges.

Extrusion coating processes are run at very high temperatures in order to maximize adhesion to the substrate. Consequently shear viscosity is relatively low particularly with LDPE and the effects of viscoelasticity are minimized, which facilitates melt distribution across the die. Experience has shown that rounded profile T slot manifold distributors work well in this process. These have the added advantage of being easier to machine and avoid leakage problems that can occur with more complex feed channel profiles cut from two blocks of steel.

Since the die controls the uniformity of the coating it must be accurately machined of hard tool steel and incorporate highly polished channels. Its lower part has a flexible V shaped cross section that can be accommodated near the nip formed by the chill roll unit. The lower portion of the die includes hardened inserts or lands. The lands build up resistance to the melt flow and develop back pressure.

To avoid corrosion with polymers such as ethylene acrylic acid copolymers (EAA) and ionomers, it is important that all surfaces in contact with the melt are duplex chrome plated or nickel plated. Chrome plated surfaces have been shown to have minimum adhesion to acid polymers. Carbon steels are not recommended.

Coating dies must be wider than the substrate to accommodate the neck-in of the melt as it emerges from the slot. The die opening (die gap) should be uniform across its width. Recommended die gap settings for LDPE are 0.5-0.6 mm. Wider die gaps may be used to reduce pressure or eliminate the formation of interlayer instabilities (zigzag waves) in coextrusions. Extrusion of polyester and polypropylene may require wider die gaps.

It is possible to extrude practically any width of coating narrower than the die by fitting deckles at both ends. Deckles are die inserts that determine the coating width from the slot die. They adjust the melt flow at the edges to the required width.

Internal and external deckles are sketched in **Figure 4.2**. Both types of deckle system are adjustable. External deckles are bolted to the die lip and simply block the ends of the die. They create dead areas behind them, which can lead to degraded material build-up. This requires periodic stopping and purging to maintain product quality.

Internal deckles make use of internal plugs, which fill space in the distributor and prevent the formation of dead areas. They are more costly and can take longer to change than external deckles. Internal deckles work well with dies of uniform distributor section (T slot).

Over the years die manufacturers have developed proprietary deckle systems that claim to drastically reduce neck-in/edge beading. Edge bead reduction systems will usually have three deckles to control the melt flow and reduce the beading effect.

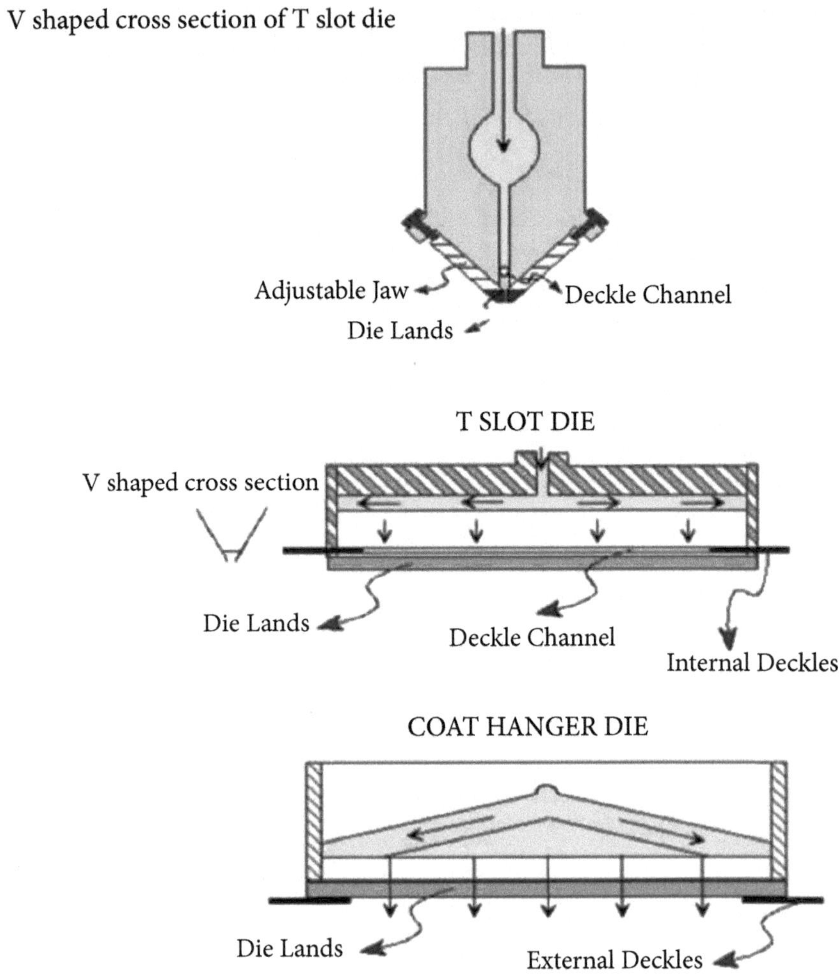

Figure 4.2

Basic coating die designs

These edge bead reduction systems are fitted with independently adjustable components that are claimed to eliminate edge trim from the process. These also allow the operator to change the melt width during operation (on-the-fly width changes). Motorized and automated systems have been developed which separately adjust the internal and external deckles, and can be set to return to the same width settings every time a particular product is processed.

Monolayer coating dies usually have a round feed channel, which is simpler to machine and easier to clean. However, for coextrusion pear or 'tear drop' shaped

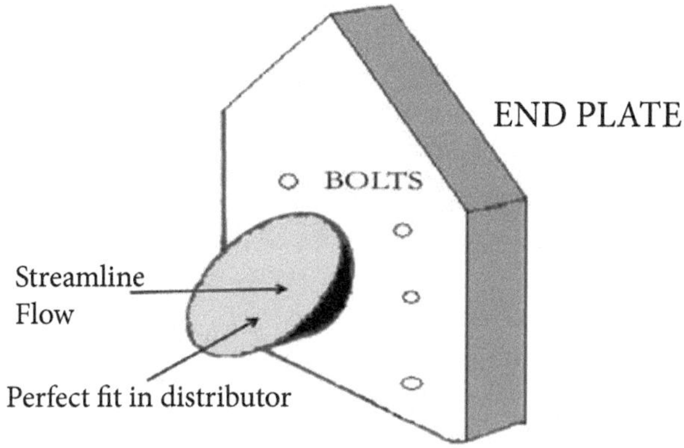

Figure 4.3

Die plate fitted at die edges

feed channels are recommended by some die manufacturers. In general with more complex multilayer flows the coat hanger die is recommended.

An important consideration in die design is the possible formation of die build-up, which can occur with all dies. This is an exit phenomenon, which results in debris from the melt attaching itself to the edge of the die lip requiring frequent machine stoppages for cleaning. Die build-up is associated with the following phenomena:

- Die swell
- Broad MWD resins
- Additives
- Oxidation
- Moisture

Build-up is more likely to form with sharp edged lips. Therefore, the recommended method to minimize this defect is to have rounded edges (6-12°) at the lip.

The presence of pigments such as titanium dioxide (TiO_2), which may initiate localized polymer degradation, can also result in build-up. Pigment masterbatch selection is, therefore, crucially important to minimize the risk of such phenomena occurring.

4.1 Automatic dies

The ideal extrusion coating die should incorporate a flexible lip that allows an easy, fast adjustment of the melt thickness. Computer aided profile control systems have made it possible to automatically fine-tune the opening and closing of the die lips. With feedback from a scanning beta gauge detector or similar device it is possible to automatically correct for thickness variations yielding coatings with a cross-directional profile of ±2% variation. **Figure 4.4** shows a basic flat die design with flexible lips. The lips are adjusted by varying the heat input into the die bolts, which determine the opening of the lips. The bolt temperatures are controlled independently with built in cartridge heaters and thermocouples. Simpler systems control temperature by varying the power (wattage) input into the heaters.

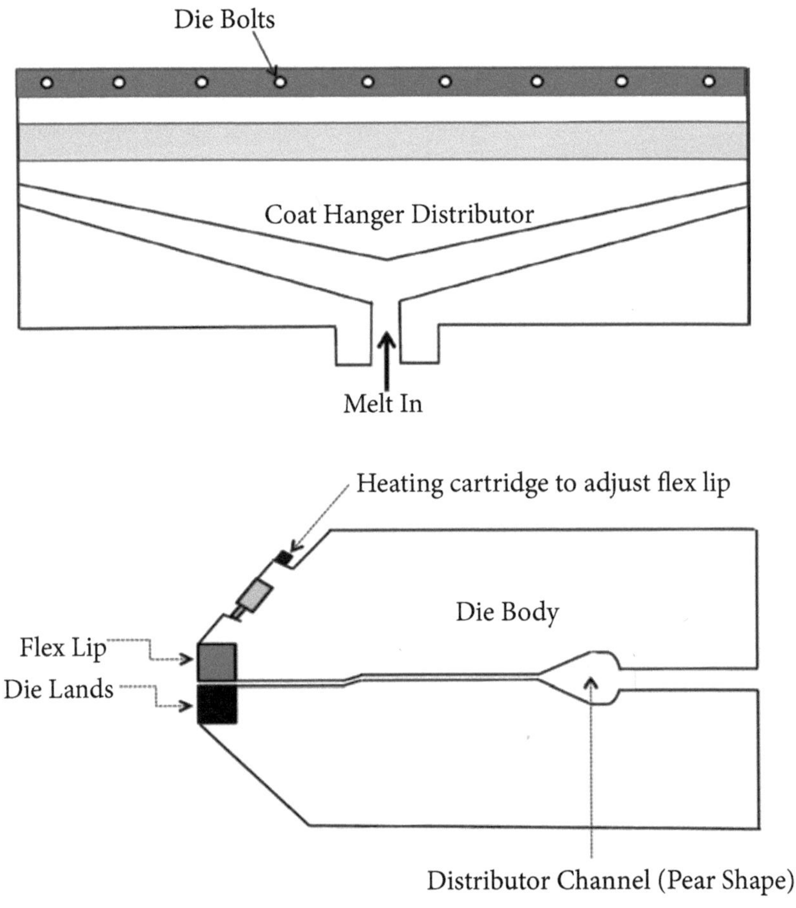

Figure 4.4

Cross section of a typical flat die

On heating, the bolts will expand and the die lips will close. On cooling the bolts contract and the lips will open from the pressure of the melt inside the die. In extrusion coating, melt viscosity is normally low and not enough internal pressure will be available to quickly force the die lips to open. Delays of twenty minutes or more are possible before the required coating thickness is delivered from the die. To accelerate the response push-pull systems are used. Spring back is faster with these designs some of which can adjust the thickness within three minutes.

Most automatic die designs incorporate permanent cooling with jets of air impinging on the bolts. Heating is practically continuous to compensate for the cooling. These controls are essential for efficient start-up and process stabilization in coextrusion.

4.2 Flat die computer simulations

Using PC simulation software a flat die was configured as in **Figure 4.5** and **Table 4.2**. The die is 1500 mm wide with a round feed channel of 30 mm diameter. The T slot distributor is 50 mm long from the middle to the edges with a constant 3 mm gap. The flow path is made up of three chambers plus the adjustable lips. A relaxation chamber, which opens out from 1 to 4 mm is included. The final lips consist of a 10 mm land with 0.6 mm gap across the 1500 mm width. The die is fed with 450 kg/hr. of polymer at a temperature of 315 °C. All die surface temperatures were also set at 315 °C. An 8.0 dg/min. melt index coating grade LDPE was entered.

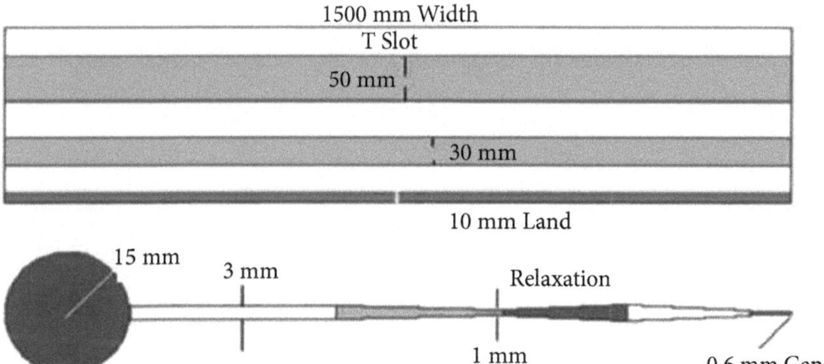

Figure 4.5

Outline of T slot die

Table 4.2
T slot flat die flow channels

Die section	Length (mm)	Width/gap (mm)	
Feed Channel: 15 mm radius		Begin	End
Distributor/restrictor	50	3	3
Chamber #1	40	3	1
Chamber #2	30	1	4
Chamber #3	30	4	1.0

Figure 4.6 shows the thickness distribution across half the die width. The variation is approximately ±3% from the middle to the edges of the die. The calculated shear stress at the lips is 25 kPa and presents no danger of surface defects. This simple T slot design works well with coating grade LDPE at high temperature and exemplifies the adoption of these designs by the extrusion coating industry.

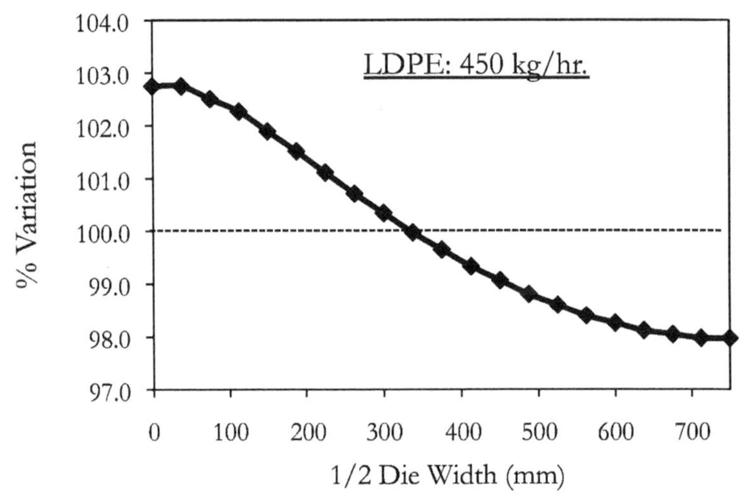

Figure 4.6

Flow distribution across die exit

4.3 Design modification of T slot die

The effect of the following die design features on melt distribution was evaluated. LDPE mass flow rate was 450 kg/hr. and all temperatures were set at 315 °C.

- Die gap (die lip opening)
- Die lip land length

- Feed channel diameter (T slot)
- Die width and residence time

Using the same design as **Figure 4.5**, die gaps were varied from 0.5 to 1.5 mm. The effects on thickness distribution were predicted by computerised simulations and the results plotted in **Figure 4.7**. The graph shows the half die width for the volumetric flow distribution at the lips. The narrowest die gap develops the lowest variation in melt distribution across the die as shown, i.e., the 0.5 mm gap is preferred.

The next step was to optimize the land length of the lips. Fixing the die gap at 0.6 mm, the effect of lands of 5 and 10 mm length were compared.

Figure 4.8 shows that the longer land of 10 mm results in a narrower flow distribution and is, therefore, preferred. The impact of the land changes on thickness distribution is less pronounced than with the die gap adjustments.

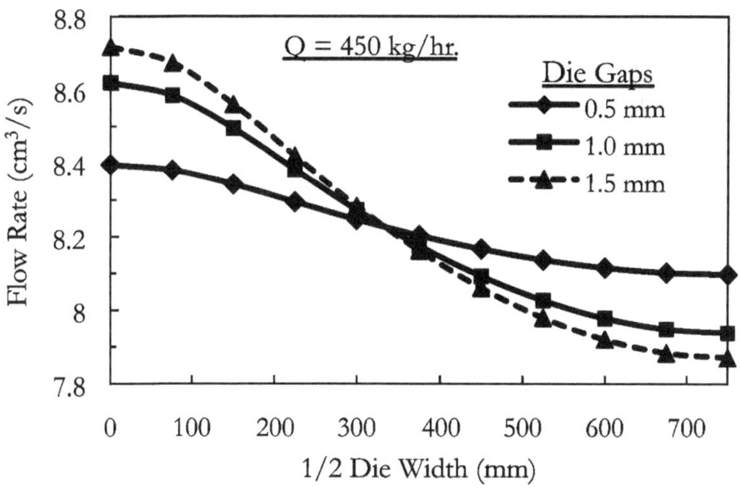

Figure 4.7

Effect of die gap on thickness distribution

The effect of die width on melt distribution was evaluated next. The 1500 mm die was extended to 3000 mm. LDPE mass flow rate was fixed at 3.33 kg/cm/hr. of width. This was in effect 500 kg/hr. for the original die and 1000 kg/hr. for the 3000 mm die. Die gap was 0.6 mm and land length 10 mm. Temperature was 315 °C.

The initial result showed the inadequacy of the 30 mm diameter feed channel to distribute the melt across the 3000 mm die. The feed channel diameter

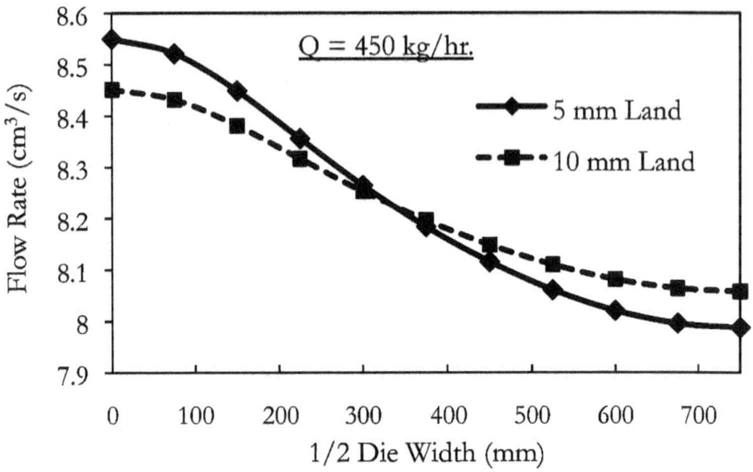

Figure 4.8

Effect of land length on thickness distribution

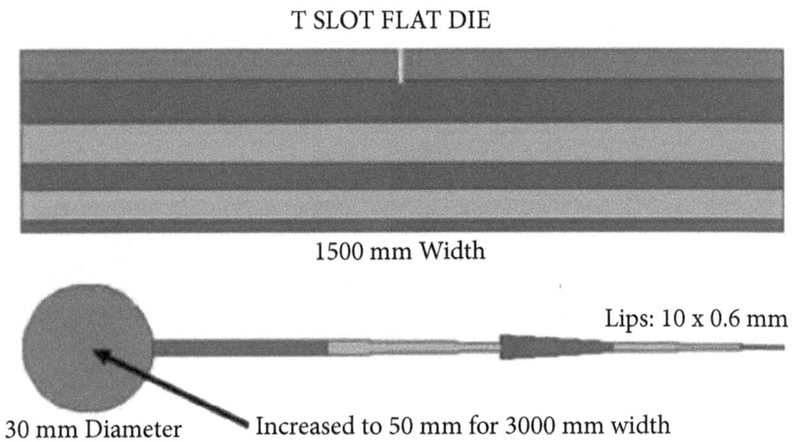

Figure 4.9

Schematic of T slot die with larger feed channel

was enlarged to 50 mm as in **Figure 4.9**. **Figure 4.10** shows the significant improvement in flow distribution in the 3000 mm wide die with the enlarged diameter feed channel. Expanding the diameter of the feed channel above 50 mm did not improve distribution. There seems to be an optimum ratio between the die width and the feed channel size.

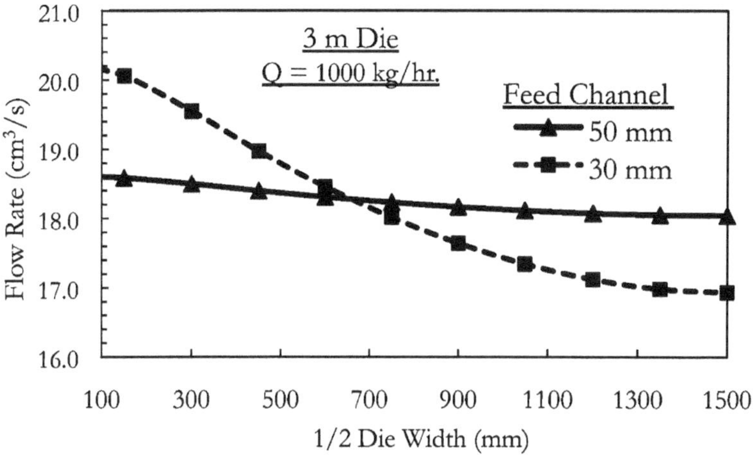

Figure 4.10

Effect of feed channel diameter on distribution

Figure 4.11 compares the flow distribution (half die width) across the 1500 and 3000 mm dies. Both dies perform well with LDPE at these process settings.

From this data the effect of die width on residence time was tested numerically. Low residence times are required with heat sensitive polymers such as EVOH, EVA and ionomers. In **Figure 4.12** three die widths are compared for residence time at the die edges. This is the predicted time for the slowest flowing segment (edges) of the melt. At low outputs the time differences are much more pronounced.

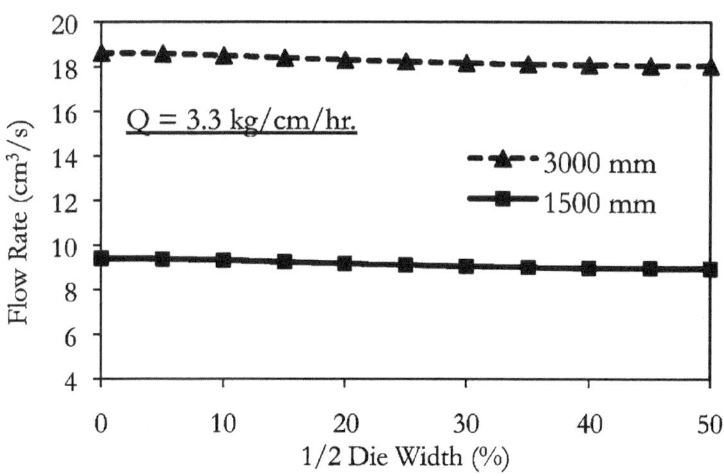

Figure 4.11

Flow distribution at 2 die widths

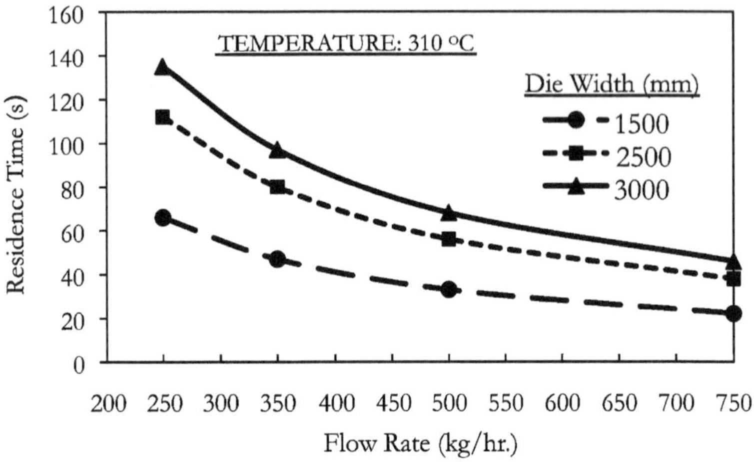

Figure 4.12

Residence times at die edges

These computer simulations demonstrate the effectiveness of the simple T slot die in processing LDPE over a wide range of commercial die widths. The effects on melt flow distribution of the die gap, land length, width and feed channel design were compared.

4.4 T slot die and other polymers

A T slot die similar to the design in **Figure 4.5** was also evaluated with the three polymers listed below:

- PP homopolymer (4.5 dg/min. melt flow rate)
- Copolyester recommended for extrusion coating (I.V. 0.75 dl/g)
- EVOH (5.5 dg/min. melt flow rate)

Recommended extrusion temperatures are as follows:

- LDPE: 315 °C
- PP: 290 °C
- Copolyester: 290 °C
- EVOH: 260 °C

The die body was set at the same temperatures as the incoming polymers.

The polypropylene was a cast film grade and not specifically recommended for extrusion coating. The shear viscosity curves of all four materials are compared in **Figure 4.13** at their suggested processing temperatures.

The LDPE at 315 °C has by far the lowest shear viscosity in the group. The PP has a much steeper slope compared to the other materials, which is significant for the flow distribution across the die. The copolyester and the EVOH have similar shaped flow curves.

The mass flow rate in the computer program is entered in weight (kg/hr.) and converted from the relevant density values into volume (cm^3/s) in order to predict velocity and pressure in the die. At a mass flow rate of 500 kg/hr. the pressure drops in the 1500 mm die with a 0.6 mm gap are shown in **Table 4.3.**

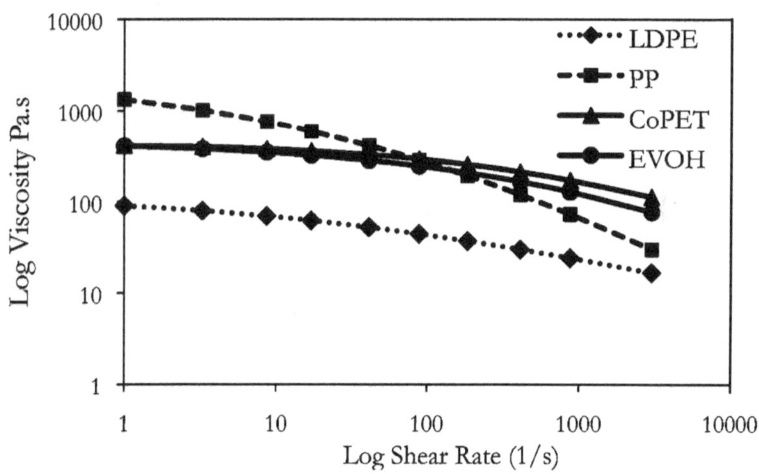

Figure 4.13

Comparative viscosity measurements for LDPE (315 °C) PP and CoPET (290 °C) EVOH (260 °C)

Table 4.3

Pressure drop in die (die gap: 0.6 mm)

Polymer	**T (°C)**	**ΔP (MPa)**
LDPE	315	0.44
PP	290	9.22
CoPET	290	12.86
EVOH	260	11.26

These measurements show the pronounced effect of polymer viscosity and process temperatures on the pressure in the die for the same throughput. To relieve these high pressures the die gap should be opened to 1.0-1.5 mm with the PP, CoPET and EVOH.

In **Figure 4.14** the flow rate variations across half of the 1500 mm die width are plotted for the four polymers. The values are expressed as percentage variations from the mean. The LDPE, CoPET and EVOH show good results with a variation of ±3% from the 100% point. The PP is significantly poorer with a variation of ±12%. Clearly this T slot die is not ideal for this particular grade of PP (note the steeper viscosity slope in **Figure 4.13**.)

Shear stresses at the die lips were also calculated for all four materials. The shear stresses at 0.6 mm die gap are plotted in **Figure 4.15**. The copolyester and the EVOH develop stresses >140 kPa indicating the possible formation of some surface defects. The LDPE and PP are well below this ceiling.

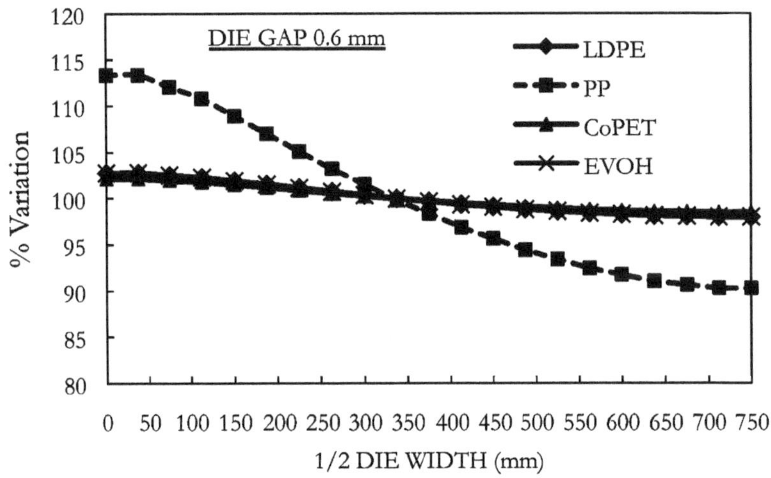

Figure 4.14

Thickness variation across the die

The die gap was now opened from 0.6 mm to 1.2 mm and the EVOH and copolyester were rerun. The shear stress predictions are shown below. All shear stress values are now below 80 kPa:

- CoPET: 77.5 kPa
- EVOH: 69.7 kPa

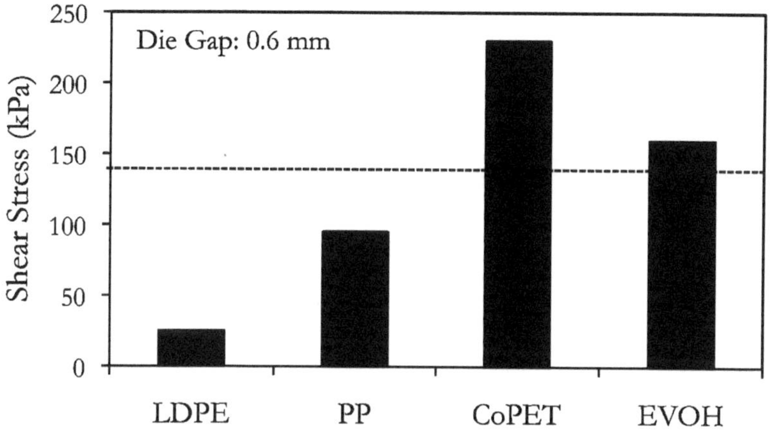

Figure 4.15

Shear stress at the die lip

Figure 4.16 compares the flow distribution for the copolyester and EVOH at the two die gaps. The poorer flow distribution control with the wider gap is clearly illustrated in these plots as they deviate from the 100% (target) position. These results again demonstrate the importance of the die gap in controlling flow distribution.

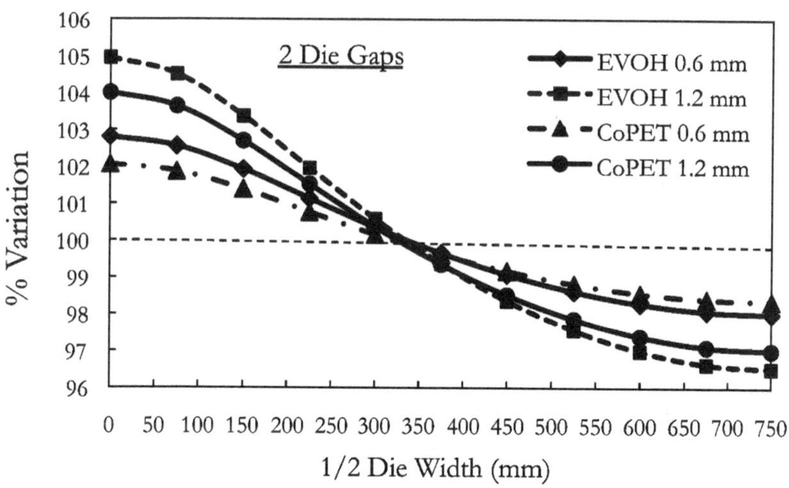

Figure 4.16

Thickness variations at 2 die gaps

The following conclusions can be drawn regarding the T slot die:

- For the best thickness distribution control always use the narrowest die gap providing shear stress and pressure are not excessive
- The T slot die works well with coating grade LDPE at high temperatures with a 0.5-0.6 mm die gap
- With PP the thickness variation is much higher. The high shear sensitivity of the PP grade used in these evaluations is adversely affecting melt distribution in the die
- The CoPET develops excessive pressure at 290 °C. It is recommended to widen the die gap to 1.5 mm to reduce pressure
- A similar outcome was observed with EVOH extruded at 260 °C
- Wide differences in pressure drop were calculated between the four polymers. This may affect die plate deflections, which will influence thickness distribution

4.5 Coat hanger die

The next step was to evaluate whether the more complex coat hanger die would provide improved thickness control particularly with the PP grade, which was unsatisfactory in the T slot design used above. A coat hanger die as shown in **Figure 4.17** was entered in the simulator. The feed channel has a pear shaped profile with a 15 mm radius. The restrictor has a constant gap of 3 mm across its width. The restrictor length forming the coat hanger is 110 mm in the middle

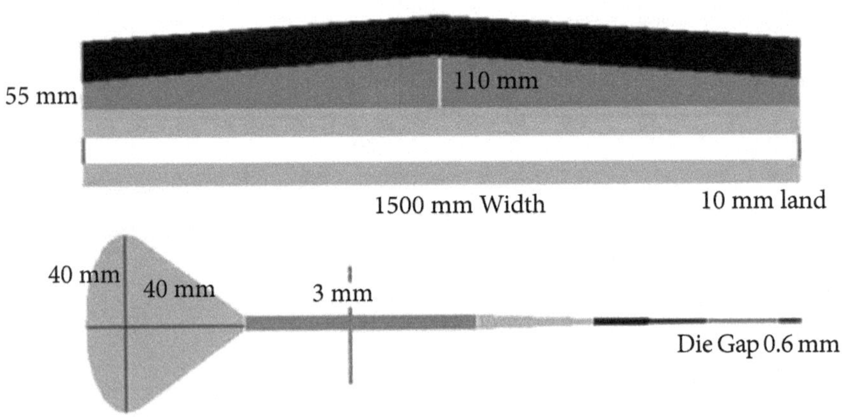

Figure 4.17

Coat hanger die

tapering down to 55 mm at the edges of the die. The lip section is identical to the T slot design used earlier. Input was 450 kg/hr. at 315 °C.

Figure 4.18 shows the much improved thickness distribution of the coat hanger die. In the T slot the LDPE melt flows preferentially down the middle and is reduced towards the edges of the die. In the coat hanger the flow is more evenly distributed across the die width. All four polymers were compared for thickness distribution in this die and plotted in **Figure 4.19**.

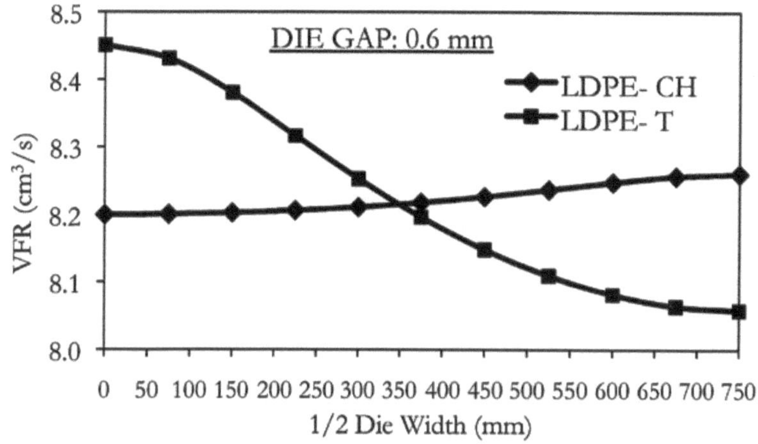

Figure 4.18

Flow distribution of T slot vs. coat hanger die

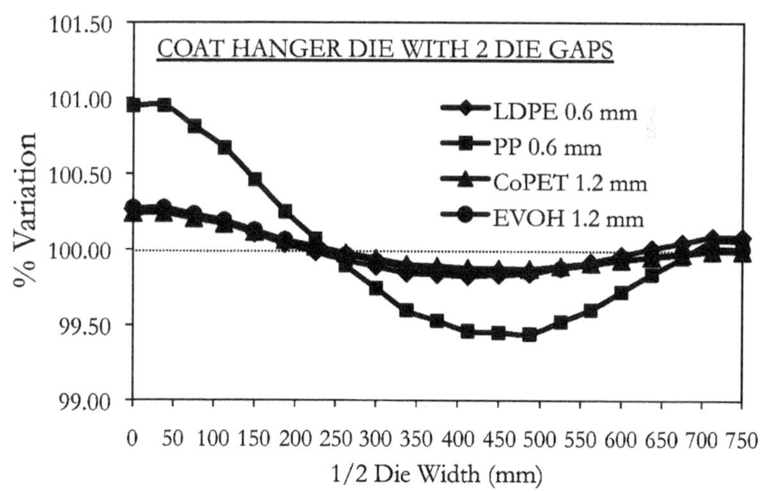

Figure 4.19

Flow distribution in coat hanger die

The mass flow rate was 450 kg/hr. set at the same temperatures as the T slot die. The copolyester and EVOH were run with a **1.2** mm die gap. The PP with a variation of ±1% shows a much improved distribution. The variations with the other three polymers are very close.

These simulations illustrate the superiority of the coat hanger die in the control of thickness distribution. The improvement is particularly significant with melts of high shear sensitivity. In coextrusion where more complex multilayer flows are involved, the coat hanger die is recommended as stated previously.

4.6 Comparative die performance

The performances of the two dies are summarized in **Table 4.4**. With the T slot die the standard deviation (s) of the PP is 0.70, which is reduced to 0.04 cm^3/s in the coat hanger die. With the die gap opened to 1.2 mm the copolyester and the EVOH have standard deviations that are close to the LDPE with 0.6 mm die gap.

Table 4.4

Statistical analysis of die performance

Polymer	Flow rate (AVG. cm^3/s)	T slot (STD (s) cm^3/s)	Coat hanger (STD (s) cm^3/s)
LDPE	8.22	0.14	0.01
PP	8.51	0.70	0.04
EVOH*	6.72	0.21	0.01
CoPET*	5.73	0.15	0.01
**1.2 mm gap*			

The T slot die is good enough with LDPE, EVOH and the CoPET. The melts spread out reasonably easily in this design at these high temperature conditions. The much more shear sensitive PP grade will benefit significantly from the coat hanger die, which is more effective in balancing melt flow across its width. The danger of excessive shear stress at the die lips has also been demonstrated. Adjusting the die gap is the simplest way of dealing with this problem if it occurs. The EVOH and copolyester grades used in these evaluations should be run at die gaps in the range of 1.2-1.5 mm. As shown elsewhere in this study, a wider die gap will reduce neck-in and is usually recommended with PP and PET, which exhibit excessive neck-in.

The intent of this study is not to recommend one type of die over another, but to demonstrate the effects of certain key design features on performance and the sensitivity of die design to different polymers and processing conditions.

Finally, it is important to remember that these computer simulations assume perfect geometry and the absence of any deformation such as 'clam shelling' that can occur during operation (effectively increasing melt flow in the centre of the die). **Figure 4.20** shows the forces developed between the plates of a simple 1500 mm wide die. According to equation 4.1, a pressure drop of 10 MPa (P) will generate an opening force (F) of 750 KN (75 tons) inside the die.

$$F = P \times S \qquad (4.1)$$

F = Force, P = Pressure, S = Surface area.

The die plates and lips will deform under such force and will require effective in-line die bolt adjustments to compensate for any flexing. Residence time behind the deckles will also cause a gradual build-up of degraded polymer at the edges, which may disturb flow in those regions. Other variables such as screw pulsation will also cause thickness variations adding to the complexity of the problem.

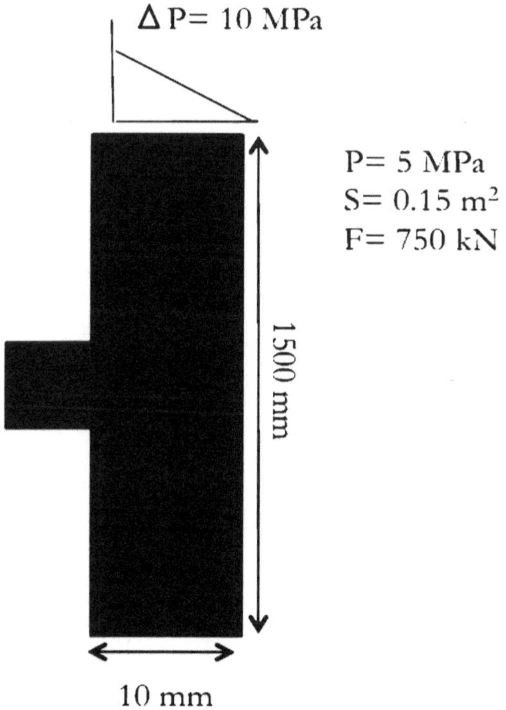

Figure 4.20

Pressure in flat die

***Note**: The computerized modelling used in these studies does not include the effects of normal and elongational forces, which will influence the pressure and flow distributions across the flow fields. Basing the calculations only on shear viscosity data will tend to predict narrower thickness distributions than the reality. However, the overall shape and trends of the flow distributions from the die models do reflect actual experimental results.*

Chapter 5

Adhesion in extrusion coating

Adhesion between the extruded polymer and the substrate is the principle criteria by which the quality of the final product will be judged. **Figure 5.1** illustrates the events taking place in the air gap as the molten polymer is drawn from the die and combines with the substrate in the coater. The melt is stretched and begins to cool and crystallize as it enters the chill roll nip. With polyethylene the principle chemical reaction is the rapid melt oxidation at high temperature in the air gap.

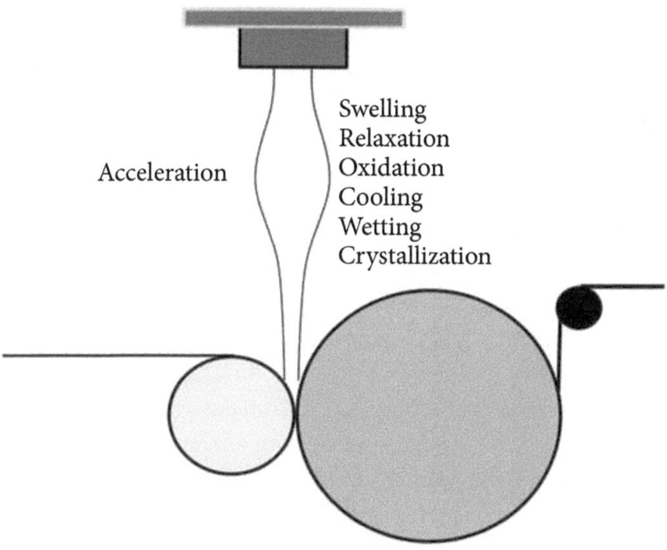

Figure 5.1

Polymer melt entering the nip assembly

In the nip the melt is pressed and spreads on the substrate and is simultaneously cooled to a tough cohesive mass. The following types of attractive forces can occur as two materials combine:

- Mechanical adhesion
- Diffusion
- Adsorption: dipole-dipole attractions

- Covalent bonds
- Hydrogen bonds

In mechanical adhesion the materials are held together by interlocking as the voids or pores are filled. Velcro is a well known type of interlocking system. Paper adhesives provide another example of mechanical bonding as the porous fibrous structure is filled by the liquid adhesive.

Bonding by diffusion occurs when the molecules of both materials are soluble in each other. This can occur with polymer chains where one end of the molecule diffuses into the adjacent material. Heat sealing and coextrusion are typical examples of this process.

Dipole-dipole forces result from **adsorption,** which holds two materials together. Attraction between molecules occurs where regions of positive and negative charges are formed. As an example the oxidation of polyethylene forms polar species, which are attracted to surfaces that have a complementary charge density. These forces range from 5-20 kJ per mole.

Much stronger bonds, 3-4 orders of magnitude higher, are formed when **covalent** linkages occur. This entails the sharing of outer electrons at the interface. The weaker **hydrogen** bonds are formed if the materials share a hydrogen nucleus. The ideal adhesion system will combine mechanical adhesion with covalent bonds.

It is important for the melt to spread and wet the substrate's surface to maximize interfacial contact. With low density polyethylene (LDPE) **oxidation** is necessary to enhance these surface interactions. Oxidation generates polar species most of which are believed to be formed in the air gap as the melt exits the die at high temperature. These consist mainly of aldehydes, ketones and acids. Oxidation like all chemical reactions is temperature and time dependent. The higher the temperature and/or the longer the exposure time to air, the higher the degree of oxidation. However, excessive temperatures can weaken adhesion as this may result in unwanted degraded material, which forms weak boundary layers at the interface.

With LDPE good adhesion is generally only possible on pulp and paper (cellulose) substrates and clean aluminium foil, unless mechanical adhesion occurs on porous surfaces. The oxidized species create sufficient interactions with fibrous cellulose and in the case of aluminium with the oxides (Al_XO_Y) embedded in its surface. On plastic films, e.g., oriented polyester (OPET) and polypropylene (BOPP) LDPE will not adhere unless a chemical primer is applied to the film. On these plastic films, ethylene **ester** copolymers are recommended. The reactive comonomer (vinyl acetate or alkyl acrylate) enhances adhesion to these materials.

5.1 Effect of polyethylene additives on adhesion

Figure 5.2 shows where the adhesion between two solids will fail due to a boundary layer with low cohesive strength. As the layers are separated cohesive failure at the boundaries will cause delamination. Materials that migrate from either the coating or the substrate are examples of boundary layers that will inhibit adhesion. Surface contamination, such as residual rolling oils on aluminium foil can significantly compromise the integrity of the interface and weaken adhesion.

The effect on adhesion caused by a series of migrating additives in LDPE coated on a fine bleached paper was evaluated. The additive contents were as follows. These compounds modify surface properties and melt flow.

- Calcium stearate (CaSt), 3000 ppm
- PE wax (Wax), 1.5%
- Polyethylene glycol (PEG), 200 ppm

Extrusion temperature was 320 °C and line speed was 180 mpm and the coating weight was 30 g/m^2. The effects of the additives on adhesion are shown in **Figure 5.3**. All three additives migrate to the polymer surface forming weak boundary layers, which significantly weaken adhesion to the paper.

The glycol (PEG) is the quickest to migrate to the surface and forms a liquid layer at the interface. The addition of calcium stearate drops adhesion by half. Other examples of release agents are erucamide and oleamide, used as slip agents to reduce surface friction. These **must** be avoided. Unintentional contaminants such as trace solvents and lubricating oils from the polymer manufacturing process must also be eliminated.

For different reasons **antioxidants**, which attenuate the oxidation rate of polyethylene, should be excluded from the product. Adding as little as 50 ppm butylated hydroxytoluene (BHT) to LDPE will drop adhesion to aluminium foil from 3.0 to <0.5 N/15 mm peel strength.

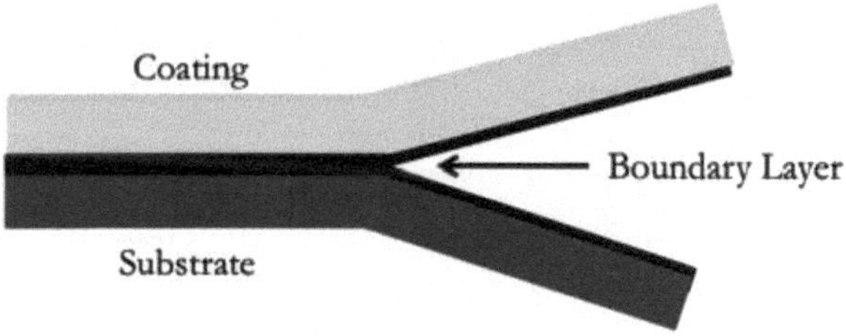

Figure 5.2

Weak boundary layer

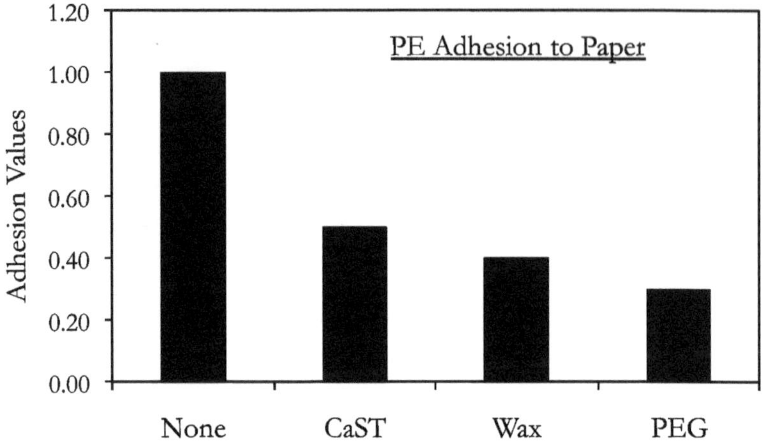

Figure 5.3

Effect of additives on adhesion

It is interesting to note that with ethylene acrylic acid copolymers (EAA), **no change** in adhesion is observed with the addition of antioxidant[1]. With these copolymers the adhesion mechanism is dominated by the chemical interaction between the acidic functionality and the substrate through other attractive forces.

It would be logical to assume that other ethylene copolymers, such as those based on vinyl acetate or alkyl acrylates, will also **not** be adversely affected by the addition of antioxidants, since adhesion is also determined by the unique interaction of the comonomer with the substrate, e.g., EVA on OPET films.

5.2 Chill roll cooling and coating weight

It is well established that high coating weights will usually improve adhesion particularly on porous paper surfaces. This will depend on the surface roughness of the substrate and the time for the melt to react at the interface. The greater the thickness the longer is the time available for the melt to penetrate and wet the surface before cooling below its freezing point. The impact of this interaction is reduced as coating weight drops below 20 g/m^2.

PC simulations were set-up to assess the elapsed time for the LDPE to solidify at the interface with the substrate. It is assumed that, once below its melting point of ~110 °C, any reaction between the LDPE and the substrate is minimal.

1 Morris, B.A. and Suzuki, N., The Case Against Oxidation as the Primary Factor for Bonding Acid Copolymers to Foil, ANTEC 2001, SPE, **2001**, p. 25-35

The software was used to predict the cooling process from the nip and along the chill roll circumference to the point where the cooled web is stripped away. The model assumes LDPE delivered at 500 kg/hr. from a 1500 mm wide flat die and coated onto 0.15 mm thick paper at a line speed of 200 mpm. The process parameters are in **Table 5.1**. The schematic of the coating model is shown in **Figure 5.4.**

Table 5.1

Cooling process parameters

Web width	1500 mm
Paper thickness	0.15 mm
Paper temperature	30 °C
Chill roll temperature	18 °C
Pressure roll temperature	45 °C
Line speed	200 mpm
LDPE melt temperature	315 °C

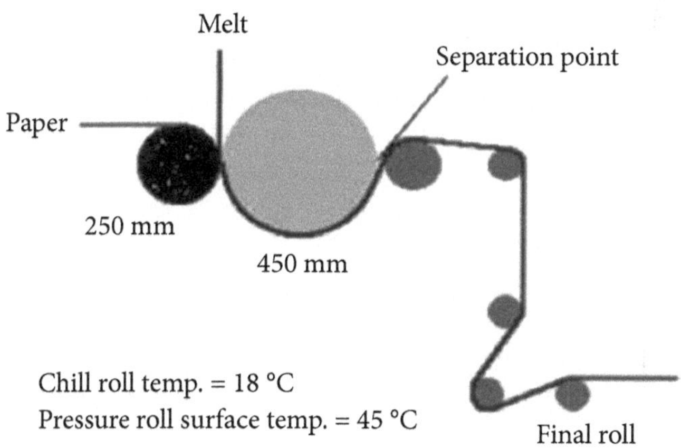

Figure 5.4

Chill roll cooling model

The melt at 315 °C enters the nip with the paper and is rapidly quenched as it contacts the chill roll set at 18 °C. The position along the chill roll for the interface temperature to drop below 100 °C is calculated. **Figure 5.5** plots the temperature profile at a cross cut exactly 25 mm from the nip. At this position the interface temperature is already well below melting point (<80 °C). The calculated elapsed time to this point is 7.5 ms (milliseconds) at 200 mpm.

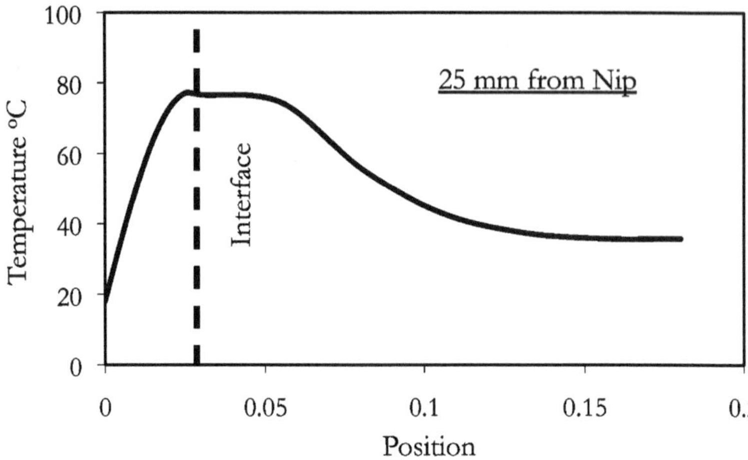

Figure 5.5

Cross cut of web temperature

The simulation was repeated with 250 and 1000 kg/hr. of LDPE entered in the model. The intent was to vary coating weight at a constant line speed. The cooling times along the chill roll to drop the temperature to 100 °C were calculated and plotted in **Figure 5.6**. The time to cool to 100 °C ranges from 2 to 21 ms over the LDPE coating weight span from 17-67 g/m^2. Any action that delays the cooling process will improve adhesion. From **Figure 5.6**, at 20 g/m^2 the cooling time is 2 ms and increases to 12 ms at 50 g/m^2. A ***six-fold* increase**. This will have a

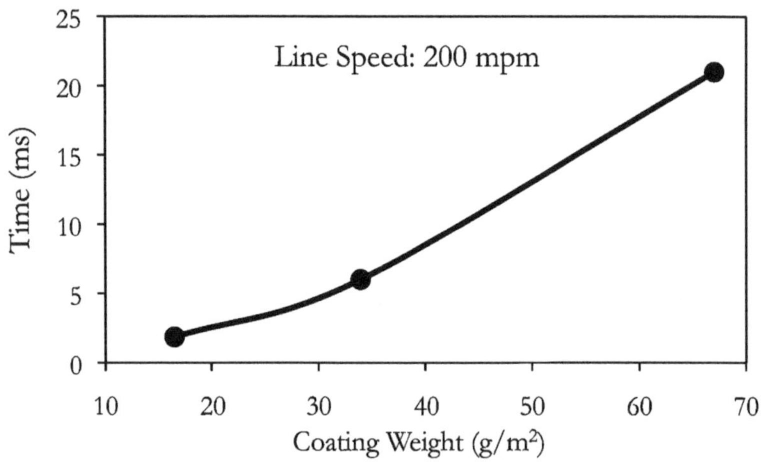

Figure 5.6

Cooling time to 100 °C at interface

considerable impact on the interactions at the interface: adhesion will improve with the longer cooling time.

Improving adhesion with increasing coating weight is demonstrated in **Figure 5.7**. LDPE extruded at 307 °C is coated on aluminium foil at 120 mpm line speed. Adhesion improves from 0.9 to 2.8 N/15 mm over the coating weight range of 10 to 40 g/m^2. Higher chill roll temperatures will also retard cooling and increase adhesion, but with the risk of unwanted chill roll sticking. Finally, retarding the cooling process will also increase melt relaxation time (see section 5.6), which will also benefit adhesion.

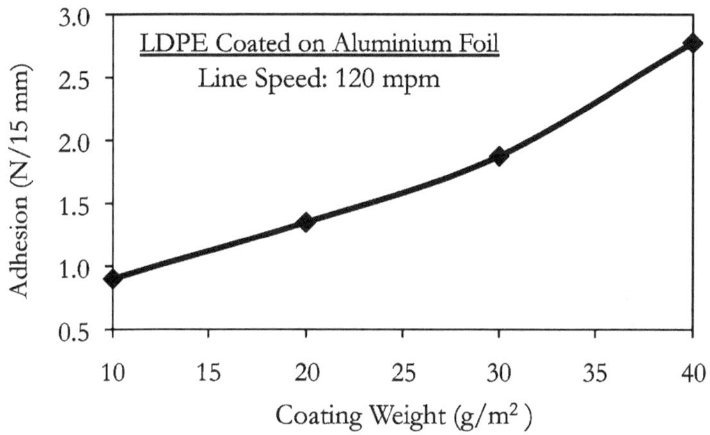

Figure 5.7

Adhesion vs. coating weight

5.3 Effect of melt temperature on adhesion

Melt temperature is the most critical extrusion parameter to affect adhesion.

All chemical reactions are temperature and time dependent. In the extrusion coating process with LDPE temperature will determine the oxidation level and in addition will affect the reaction rate between the oxidized species and the substrate. As an example with aluminium foil high temperature is required not only to induce sufficient oxidation, but also to accelerate the reaction with the oxides on the foil's surface. With copolymers, such as ethylene acrylic acid copolymer, high temperature is still necessary to facilitate the reaction between the acid and the aluminium oxides. As will be seen later, stress relaxation will also play a rôle in determining adhesion. High temperature reduces elongational viscosity, which in turn reduces stress in the melt.

Figure 5.8 shows the impact of melt temperature changes on adhesion of LDPE coated onto the aluminium side of a paper/foil laminate. Line speed was constant. Adhesion increases exponentially over a relatively narrow temperature range. At these process conditions the oxidation rate below 300 °C is low with no measurable adhesion. At 310 °C oxidation has reached a level where adhesion rises rapidly to 3.6 N/15 mm. Adhesion reaches a plateau as the temperature is increased above 320 °C. This data shows that in this case there is little benefit in using temperatures above 320 °C.

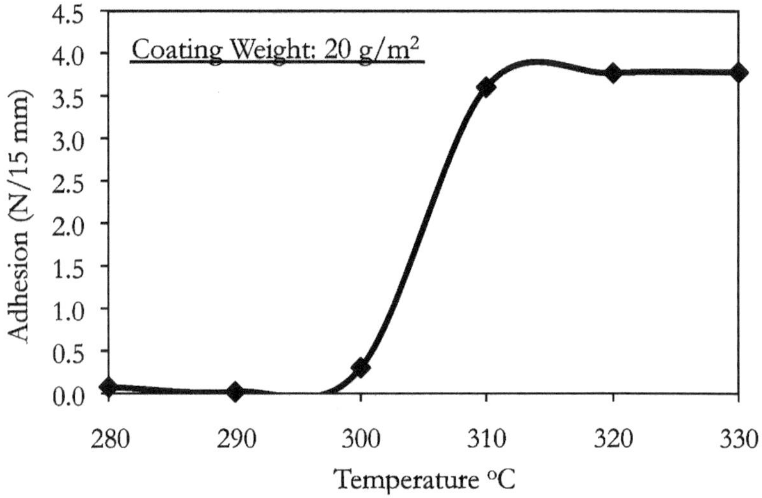

Figure 5.8

Effect of temperature on adhesion to aluminium foil

Excessive melt temperature will develop unwanted smoke, discoloration, and odour. The challenge for the extrusion coater is to use operating temperatures that optimize adhesion without *excessive* oxidation.

Figure 5.9 compares the adhesion of LDPE on two substrates at 295 and 315 °C. The plots show the significant improvement in adhesion as the temperature rises above 310 °C.

As a generalization, with LDPE, extrusion temperatures must be in the range of 310-320 °C. Lower temperatures are only feasible with oxidation boosters, such as ozonizers or very high coating weights on porous substrates. It is crucial to remember that the degree of oxidation is both temperature and time dependent.

Ethylene copolymers can enable the use of **lower** temperatures. Adhesion will depend on the specific chemical characteristics of the comonomer in relation to the substrate.

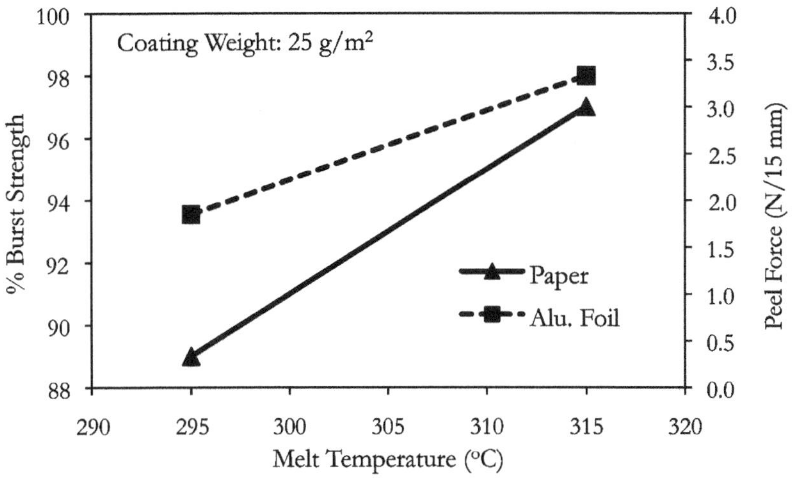

Figure 5.9

Effect of 2 temperatures on adhesion

Examples of ethylene copolymers are:

- EVA (vinyl acetate)
- EAA (acrylic acid)
- EMA (methyl acrylate)
- Other acrylates (EEA, EnBA)
- Maleic anhydride (MAH) grafted polyolefins

EAA copolymers are very effective on polar substrates, e.g., aluminium foil, paper, and metallized films. The carboxylic acid can form covalent bonds with the oxidized compounds (Al_XO_Y) embedded on the metal surface. These compounds are amphoteric and can react as bases with acrylic acid. The excellent adhesion of EAAs to pulp and paper substrates is believed to be due to hydrogen bonds formed with cellulose (-OH). EAAs are normally extruded at 285-295 °C. Odour and smoke are low in this temperature range.

EMAs adhere to non-polar materials such as polypropylene, polyester, polystyrene and PVDC coatings. They are usually processed at 305-310 °C.

The usage of EVAs is restricted by their poor heat stability above 260 °C and possible corrosion from the release of acetic acid. These copolymers are also prone to crosslinking at high temperature. However, EVAs containing 16-18% VA are extrusion coated on OPET films for the production of thermal films.

Special chill rolls are required to minimize sticking and blocking. These EVAs melt at low temperature and when heated will adhere to a variety of substrates.

The maleic anhydride grafted polyolefins involve a two step manufacturing process, which adds considerable cost compared to the production of copolymers in LDPE reactors. These products are widely used as thin adhesive layers in coextrusion. They enable the coextrusion of polar with non-polar polymers, e.g., polyamide with polyethylene.

5.4 Time in air gap

The air gap is the distance from the die exit to the nip where the chill roll is in contact with the rubber pressure roll. This zone is shown schematically in **Figure 5.10.**

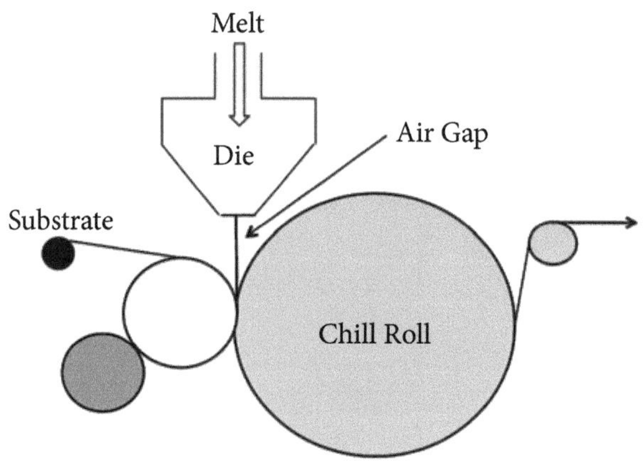

Figure 5.10

The air gap

The time that the melt resides in the air gap will determine the degree of oxidation and cooling that takes place as it contacts the substrate. An early observation of this phenomenon is reported in **Table 5.2**. In this experiment a 250 g/m^2 semi-bleached board was coated with an 8 dg/min. melt index LDPE at 315 °C at a line speed of 300 mpm. Adhesion to the board was measured by a method that determines the pneumatic pressure required to lift the LDPE coating off the paper. **Table 5.2** shows adhesion improving progressively as the air gap is increased from 100 to 200 mm. Fibre tear is achieved at 200 mm (8″) at a line speed of 300 mpm.

Table 5.2

Effect of air gap on adhesion (coating weight 40 g/m²)

Air gap (mm) 300 mpm	Adhesion (Bar)
100	1
150	1.5
200	2.5/fibre tear

Figure 5.11 plots line speed versus LDPE adhesion to an aluminium foil substrate. Adhesion falls rapidly in the between 100 and 200 mpm. Line speed controls the time in the air gap, which controls oxidation.

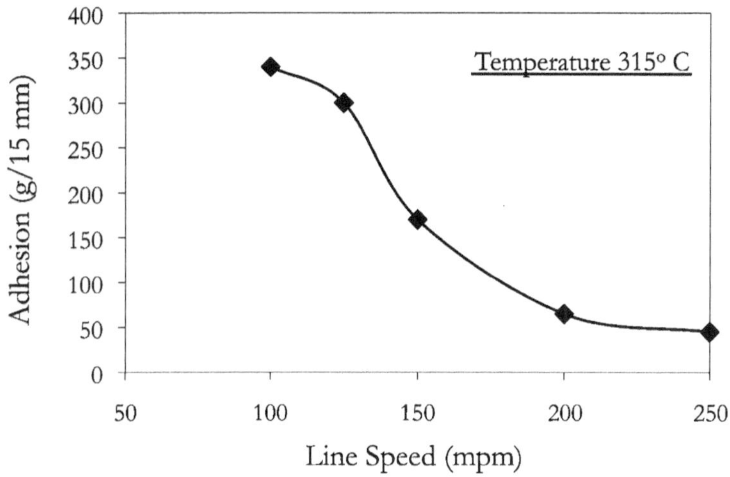

Figure 5.11

Effect of line speed on adhesion (LDPE @ 40 g/m²)

Figure 5.12 shows similar data at 305 and 310 °C also on foil. This graph shows the combined effects of temperature and line speed on adhesion.

These observations demonstrate that at any given temperature: both the length of the air gap and line speed will have a significant effect on adhesion. Both these process parameters govern the **time in the air gap**.

To calculate the time in the air gap, the average melt velocity must be known. **Figure 5.13** plots the computerised numerical values of velocity in a 160 mm air gap at a line speed of 150 mpm. The average velocity from this data is 44.7 mpm. The result is a time in air gap of 215 ms. Without the benefit of a calculator, the relationship can be simplified by utilizing the line speed as the velocity and assuming that 100 ms is the optimum time in the air gap. See Antonov and Souter[2] for further insight.

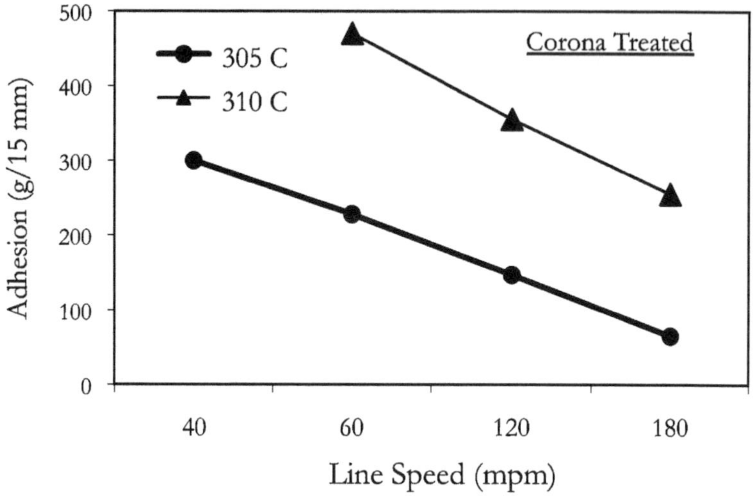

Figure 5.12

Effect of temperature and line speed on adhesion

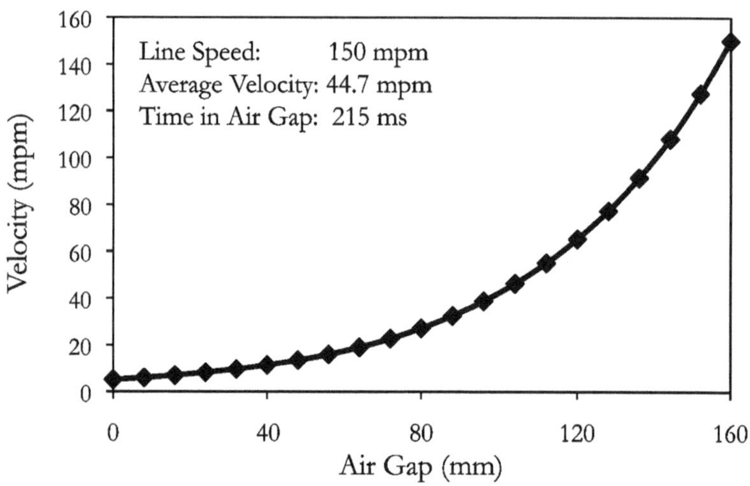

Figure 5.13

Velocity in 160 mm air gap

2 Antonov, V. and Soutar, A.M., Foil Adhesion with Copolymers: Time in the Air Gap, TAPPI Polymers, Laminations and Coatings Conference, Atlanta, TAPPI Press, **1991**, p. 553

Predicting the air gap is now reduced to a simple calculation as shown in equation 5.1. The air gap setting can be estimated for any given line speed. This is an approximation and should be used with some caution.

$$\text{Air gap (mm)} = \frac{100\ (\text{ms}) \times \text{Line speed (mpm)}}{60} \tag{5.1}$$

N.B. In the writer's opinion the 100 ms time in air gap requirement is excessive and a better approximation would be ~70 ms.

It is important to bear in mind that neck-in will increase with increasing air gap at any given line speed. Therefore, avoid excessive air gaps by fine-tuning all of these variables.

5.5 Temperature drop in the air gap

Extrusion coaters are usually reluctant to use long air gaps because of the fear of reduced adhesion caused by an excessive temperature drop as the melt travels in the air gap.

Table 5.3 describes two extrusion coating set-ups. An LDPE coating of 38 g/m^2 is applied at 100 and 200 mpm from a die with exit temperature of 315 °C. The air gap is 160 mm (6"). The assumption is that adhesion **will drop** as the line speed is doubled.

In **Figure 5.14** at 100 mpm the melt temperature entering the nip is 304 °C. At 200 mpm the time in air gap is reduced resulting in a higher temperature of

Table 5.3

Process conditions

Parameters	Conditions/process	
Melt Temperature	315 °C	
Die width	1500 mm	
Die gap	0.6 mm	
Coating weight (avg.)	38 g/m^2	
Air gap	160 mm (6")	
Air temperature round air gap	50 °C (area between hot die and nip)	
	Process #1	**Process #2**
Output	500 kg/hr.	1000 kg/hr.
Line speed	100 mpm	200 mpm

311 °C. With the temperature increased by 7 °C adhesion will still drop. The higher temperature at the nip does not compensate for the reduced oxidation time in the air gap. The slightly higher melt temperature at 200 mpm is due to the increased viscous dissipation in the die, caused by doubling the machine throughput.

In **Figure 5.15** the air gap for the 200 mpm line speed is increased to 320 mm (doubled) to equalize the time in the gap. The two melt temperatures at the nip

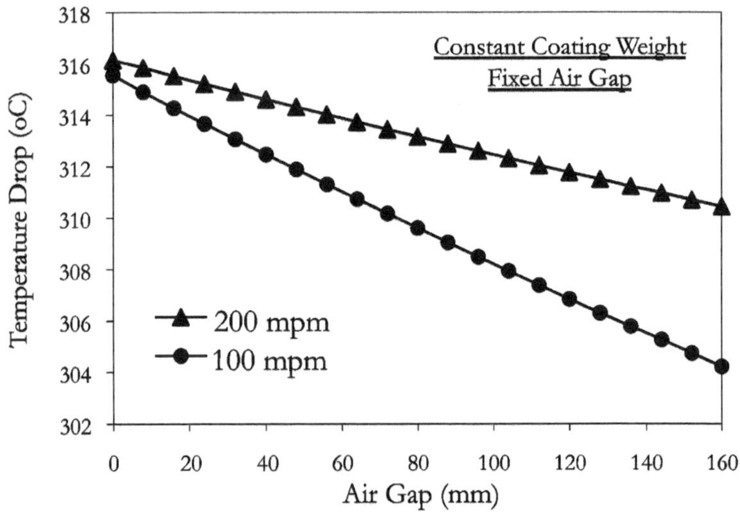

Figure 5.14

Temperature drop at 2 line speeds

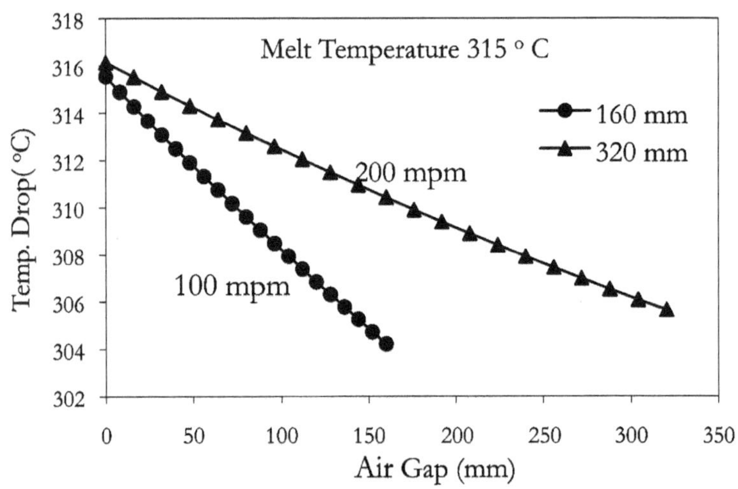

Figure 5.15

Temperature drop at 2 air gaps and line speeds

are now approximately identical and LDPE oxidation must be equal at both conditions, i.e., time and temperature drop are equalized. Adhesion should now be the same at both line speeds.

The commercial imperative is to operate the process at the highest possible line speed. Line speeds below 150 mpm are usually not economical with this capital intensive process. Precise air gap setting is the first option to evaluate in attempting to maximize adhesion at high line speed. This is the simplest to test and the least complex of the parameters that control adhesion.

Caution. It is important to be aware that the above conclusions are valid when **all other process parameters remain constant**. The calorific value of the melt will increase proportionately with the machine throughput, which may increase chill roll sticking.

Another consideration is the discharge energy from the corona treater, **if in use**. As the substrate throughput under the electrodes increases this should be compensated for by a proportional increase in power from the treater.

5.6 Elongational stress and melt relaxation

As the melt is drawn in the air gap the applied stress and strain rate will vary according to the process conditions and the polymer's elongational viscosity. It is reasonable to assume that the applied stress and relaxation time in the air gap may affect the interaction between the cooling melt and the substrate's surface thus influencing adhesion. The melt stretching regimen is a function of the molecular characteristics of the polymer, process settings, temperature and time. As an example, in a long air gap relaxation time is high, and adhesion may improve over that of a shorter air gap. This approach presents a complementary method of adding insight to the adhesion phenomena in extrusion coating. A paper by Morris discusses this concept[3]. The data in **Figure 5.16** is extracted from this paper.

The stress/strain and relaxation time induced in the extruded melt can be varied by controlling a number of parameters. These include temperature, melt velocity, air gap and die gap for a given material. A series of experiments investigating the correlation between adhesion and the stretching response of the melt were presented. To minimize the impact of oxidation an EAA copolymer (9.5% AA) extruded at approximately 290 °C was used. This resin was coated onto 50 μm

3 Morris, B.A., The Influence of Stress on Peel Strength of Acid Copolymers to Foil, Antec 1999, SPE, **1999**, p. 34

aluminium foil at a coating thickness of 25 μm. Although the effects of oxidation cannot be completely ignored its influence on adhesion at the low extrusion temperatures used in these experiments is assumed to be **negligible**.

Figure 5.16 plots the adhesion values as a function of time in air gap for the melt coated on the aluminium foil. Time was varied with both velocity (line speed) and air gap length using three die gaps to vary draw ratios. Although there is a scatter in the results the trend shows improving adhesion with increasing time in the air gap. The scatter can be partly explained by the different strain rates and melt temperature variations caused by extruder output adjustments and die gap changes. Carrying out extrusion experiments under rigorous isothermal conditions is challenging over a wide range of process variations.

With this copolymer the bond strength is primarily determined by the presence of acrylic acid. However, there is clearly a secondary effect related to the stretching regimen of the melt.

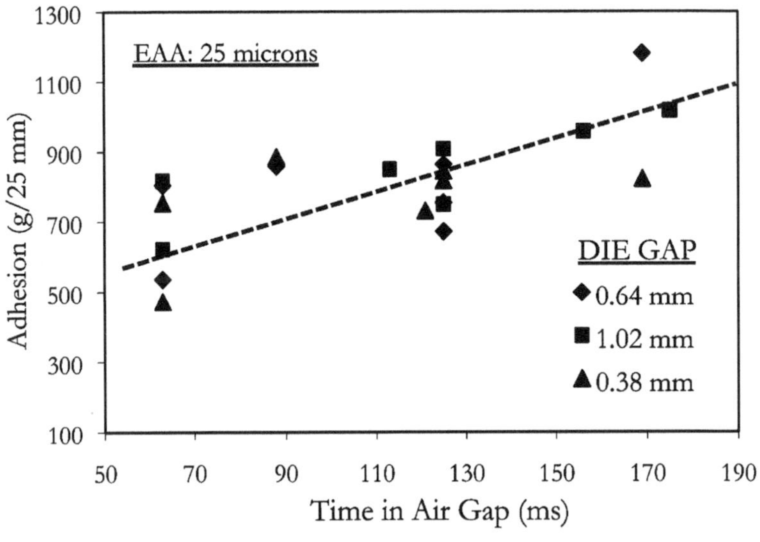

Figure 5.16

Adhesion vs. time in air gap for EAA on aluminium foil

These experiments cannot be duplicated with LDPE since at any given temperature adhesion is determined principally by **oxidation**, which is controlled by air gap and velocity, i.e. **time in air gap**.

Another interesting observation was made in comparing adhesion data derived from extrusion coatings made with blends of HDPE with LDPE. The basic experimental data is in **Table 5.4**. The substrate was a specialty paper coated at a melt temperature of 325 °C. The blend contained 35% HDPE. The adhesion results are shown in **Figure 5.17**.

Table 5.4
LD/HDPE blend experiments

	MI (dg/min)	Density (g/cm³)
HDPE (35%)	8.0	0.960
LDPE (65%)	3.5	0.923
Melt Temperature		325 °C
Coating Weight		30 g/m²
Bleached Paper Substrate		

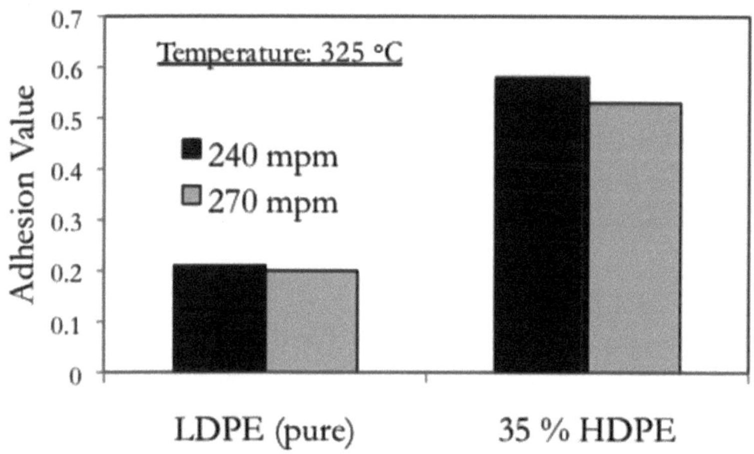

Figure 5.17
Adhesion of LD/HDPE blend on paper

These counter-intuitive results show a very significant improvement in adhesion at both line speeds in the presence of 35% HDPE in the blend. At first this outcome was attributed to the much improved chill roll release observed in the presence of the more crystalline HDPE. However, another explanation can be attributed to the strain softening effect of the HDPE. Neck-in was higher than with the pure LDPE indicating lower stress build-up resulting in an enhanced interaction on the substrate's surface. It is probable that the improved adhesion is a combination of both improved chill roll release and lower melt strength (lower stress) in this example.

Experiments based on blends of LDPE with LLDPE also indicate that adhesion can be improved by the addition of the linear polyethylene. In **Figure 5.18** adhesion results on aluminium foil at two line speeds are shown. LLDPE blends (33% b/w) of 7.0 and 20.0 dg/min. melt index were used. Melt temperature was in the range of 315-320 °C.

Adhesion is significantly improved (**average 51%**) in the presence of LLDPE. Neck-in is doubled with 33% LLDPE indicating **lower** elongational stress. However, LLDPE, particularly the lower melt index grade, will develop more viscous dissipation (higher temperature) in the extruder for a given screw speed, which may also lead to improved adhesion. But it seems unlikely that these small differences in temperature would account for 51% higher adhesion. In **Figure 5.8** (section 5.3) the adhesion curve plateaus as the melt temperature begins to increase above 315 °C. Experiments with ozonizers show no improvement in adhesion as the melt temperature increases above 305 °C. These observations are consistent with the concept that once the surface oxidation level is optimized no further improvement in adhesion is possible with polyethylene. It is therefore likely that the lower melt strength propagated by the LLDPE is having some effect. These interactions are clearly complex and delineating between the various process and material variables is not simple.

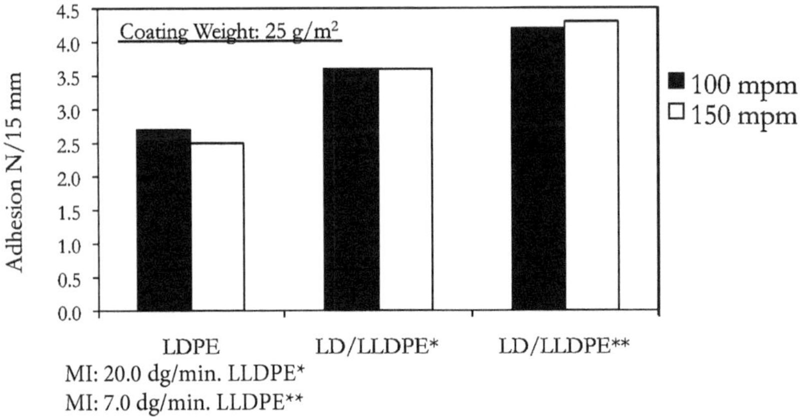

Figure 5.18

Adhesion of LD/LLDPE blends on aluminium foil

5.7 Screw speed and residence time

Polyethylene pellets as they feed into the extruder will contain some entrapped oxygen that may decompose the polymer in a very hot machine. This will result in the formation of oxidized species, which may affect adhesion and odour. These oxidation reactions are dependent on temperature and residence time in the extruder. Oxidation may also vary across the die particularly at the edges where residence time near the deckles is highest.

Screw speed determines the polymer's residence time and viscous dissipation in the extruder and may have some effect on adhesion. This phenomenon was tested with simple experiments, which involved extrusion coating an LDPE on a semi-bleached paperboard at constant temperature and line speed, but with **varying screw speed.** The surface oxidation of the coatings was evaluated by applying a calibrated ink solution on the surfaces, allowing the ink to dry then stripping it off with 1.5 cm wide adhesive tape. The residual ink retained on the LDPE coatings was used as a qualitative measure of the surface oxidation.

In **Figure 5.19** the melt temperature at the die exit was maintained at 284 °C and the screw speed was varied from 37-142 rpm in a 114 mm (4½") diameter extruder. As can be seen ink retention is reduced with increasing screw speed. This indicates that as residence time in the extruder is diminished oxidation from entrapped air in the machine drops leading to reduced ink adhesion.

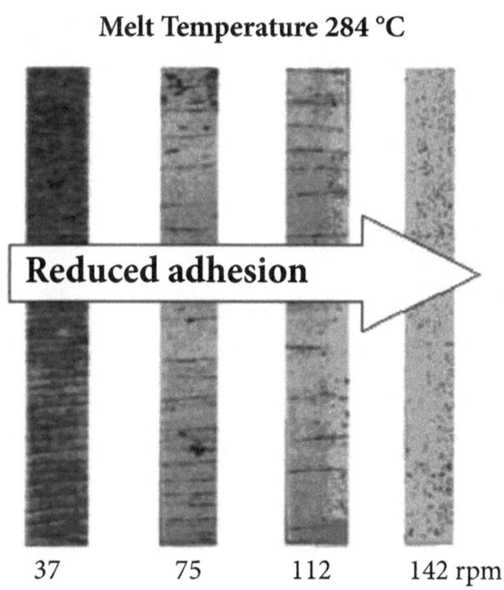

Figure 5.19

Effect of screw speed on LDPE surface oxidation

This conclusion raises some interesting issues regarding sizing of extruders. A large extruder operating at comparatively low screw speed will result in a high residence time compared to a smaller machine running at higher screw speed for the same output. Adhesion may, therefore, differ for the same barrel set temperatures, line speed and coating weight. If adhesion is not significantly affected it is possible that organoleptic properties, such as odour are altered. Using PC extrusion simulation software the residence times in 114 mm and 150 mm diameter extruders were compared for a fixed LDPE output of 350 kg/hr. Both extruders have an LD ratio of 28 : 1 and the same screw compression ratios. The results are in **Table 5.5**.

Table 5.5

Residence times in 2 extruders with LD ratio 28 : 1

Extruder Size	Screw Speed (rpm)	Output (kg/hr.)	Res. Time (s)
114 mm (4½″)	154	351	153
150 mm (6″)	108	350	212

For approximately 350 kg/hr. output the residence times for the two extruders are 153 and 212 seconds respectively. This difference may be significant. There is much anecdotal evidence to suggest that extrusion coaters can have difficulties replicating results with machines of different sizes. The dangerous assumption is that with a fixed set of process conditions the outcome must be constant. This may not always be the case. Measuring instruments will not always detect subtle differences between extruder designs, which may add to the difficulties in analyzing and detecting the source of product defects. An important variable is the poor mixing capability of single screw machines that can develop temperature gradients of some 20 °C from the barrel wall to screw root unless efficient mixing devices are incorporated.

The following conclusions regarding adhesion in the extrusion coating process can be made.

Primary effect. With polyethylene adhesion is primarily controlled by the degree of **oxidation** and/or the presence of a **comonomer** that interacts with the substrate. The degree of oxidation is a function of temperature and time. The comonomers are typically either acids (acrylic) or esters (vinyl acetate or alkyl acrylates). The acid copolymers (EAA) are the most versatile as they adhere to metals, and pulp and paper substrates. The esters can be used with non-polar plastic films (OPET, BOPP etc.)

Secondary effect. Melt elongational stress and relaxation time will also play a significant role in the development of adhesion as the melt is quenched and interacts with the substrate. This is particularly significant at reduced temperatures when the elongational stress can be high.

Time in the air gap governs both oxidation and stress relaxation time. Consequently, with a given material, coating weight and temperature this becomes a key parameter in the adhesion mechanism.

Other factors which must all be carefully considered in the drive to optimize the process include the following:

- Time in extruder (screw speed and size)
- Chill roll cooling

Chapter 6

Adhesion promotion methods

The following methods to improve adhesion between the polymeric melt and the substrate or film are reviewed in this chapter:

- Corona treatment
- High density atmospheric plasma
- Ozone treatment of the melt
- Flame treatment
- Chemical priming

The ability of a liquid to wet a surface will depend on the surface energies of the liquid and substrate. A material is wetted if its surface energy is higher than that of the applied liquid. To achieve sufficient interaction the surface tension of the film must be at least 10 mN/m higher than the applied adhesive. All the techniques listed above are intended to modify the surface properties of the interacting surfaces.

Table 6.1 shows the critical surface tensions of a variety of plastic materials.

Table 6.1
Critical surface tensions

Material	Surface tension (mN/m)
Polypropylene	29
Polyethylene	31
Polystyrene	32
Ionomer	33
PVDC	40
Cellulose (RCF)	42
Polyester	43
Polyamide	46
EVOH	37
Acrylic/acrylate	39

The polyolefins have the lowest surface tensions. Polypropylene is one of the most difficult materials to print and laminate.

6.1 Corona treatment

Corona discharge treatment has for many years been an important weapon in the armoury of film converters and extrusion coaters. The technique is widely used to prepare surfaces for printing, laminating and coating.

Corona is generated by applying a high frequency, high voltage current to an electrode separated from an earthed surface by an air gap, a substrate and a layer of dielectric material. Frequencies are in the range of 10 to 50 kHz with peak voltages of 30 kV. When the voltage across the gap reaches 3,000 to 5,000 volts per millimetre (mm) the free electrons accelerate towards the positive electrode with energy high enough to displace electrons from molecules in the air gap. The outcome is a cascade effect with electrons and ions being produced, which results in a flow of current across the gap. Any variance that occurs is due to surrounding conditions such as humidity. Large ionization currents are produced and as the power input is increased the potential across the air gap remains more or less constant. This means that increasing power across the corona will proportionately accelerate the rate at which the charged particles flow.

For efficient treatment switch mode power systems are used. These provide a quantitative measure of the match between the load and the corona power. These switch mode systems makes it possible to install display panels that show the operators true output conditions.

When a dielectric material such as a plastic film is placed in the path of these particles, surface changes will take place on the plastic. The accelerated electrons on the surface of the film will cause the polymeric molecules to rupture and form reactive sites. The ozone generated by the discharge will oxidize these sites thus increasing surface tension. As little as 0.1 μm of the surface is modified, meaning that the basic properties of the substrate or film are unaffected by this treatment.

The electrode or applicator consists of one or more metal electrodes spaced round the circumference of a metal roller covered with a dielectric. The dielectric sleeve must be made from an ozone resistant material and be completely free of cavities or micro-holes, which may cause breakdown during operation. The material must also have a low dielectric loss to minimize heating. Roll covers can be made from vulcanized elastomers or ceramic materials. The ceramics are less prone to decomposition from the discharge energy than organic elastomers.

The optimum air gap depending on the dielectric material and the frequency is in the range of 1-2.5 mm. Electrode design is critical. Treating efficiency is improved by the use of multi-electrode configurations to spread the corona discharge over the widest area.

In extrusion coating experience has shown that the electrode assembly must be placed as close as possible to the die/air gap for maximum efficiency. It is not clear why! A possible explanation is that the shorter the distance the higher the 'activated oxygen' concentration entering the nip to accelerate polymer oxidation.

With a conductive material such as aluminium foil the electrode system is reversed. The foil is maintained at an earth potential by passing it over a non-insulated metal roller covered with a dielectric.

Recommended generator sizes are listed in **Table 6.2**. The power values are **net** power at the electrode and **not** input power. As mentioned above state-of-the-art generators include switch mode power systems that provide a quantitative measure of the match between the load and corona power source. This enables power supply display readings that show the operator true output conditions.

Table 6.2

Power requirements

Substrate	Power (W/m²/min)
Paper (20 g/m²)	15-20
Paper/board (>20 g/m²)	30-40
Al foil	25-30
PE film treatment	20-25

Figure 6.1 shows the effect on adhesion of increased power from the corona generator. These extrusion coating experiments were carried out at 320 °C, at a line speed of 280 mpm and coating weight of 30 g/m². The effective die width was 55 mm. A highly calendared paper substrate was used.

Figure 6.1 clearly shows that as the input power increases adhesion improves progressively from poor to partial fibre tear and ultimately to full fibre tear. At this point the bond with LDPE is greater than the cohesive strength of the paper. In this example the S curve peaks and plateaus at 5 kW.

The treat intensity to reach optimum adhesion for this experiment can be calculated from equation 6.1. The generator setting of 5 kW is equivalent to 6 kVA.

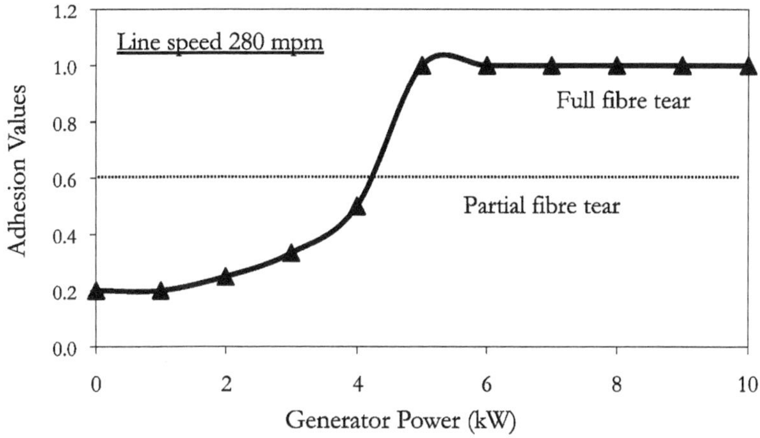

Figure 6.1

Effect of corona treatment on adhesion

$$T = \frac{p \times 1000}{w \times v} \qquad (6.1)$$

p= effective power (kVA)
w= web width (m)
v= velocity (mpm)
T= treat intensity (W/m^2/min.)

If w = 0.55m and v = 280mpm, then the calculated treat intensity will be 38.96 W/m^2/min. In these calculations it is important to bear in mind the surface characteristics of the substrate. A coarse and fibrous surface will make it easier for the melt to adhere to it.

Corona treatment is easy and cheap to use and works well on pulp and paper products, plastic films and aluminium foil. With aluminium foil bearing substrates LDPE adhesion will generally be acceptable provided that non-aggressive products are packed. For more demanding product resistance, EAA copolymers or zinc ionomers should be used.

6.2 High density atmospheric plasma

Recently new techniques have been developed to improve the efficiency of film laminating processes. High density atmospheric plasma represents a newer generation of surface treatment that is claimed to be ideal for film lamination.

Atmospheric plasma involves the electrical ionization of a gas, similar to corona treatment. The plasma creates a cloud of ionized gas that reacts with the substrate surface to raise its surface energy and improve adhesion. The process operates at lower voltage levels than corona discharge. An AC power source is used with a proprietary electrode, a dielectric between the electrodes and a selected gas mixture as the plasma medium. Helium and acetylene are suggested gas mixtures for plastic film applications. Other gases, such as oxygen (O_2) nitrogen (N_2) and carbon dioxide (CO_2) can also be used.

With nitrogen reactive nitrile groups are formed, which can form strong bonding sites. The absence of oxygen prevents oxidation and chain cleavage. An interesting claim is that the equipment can be adapted to deposit ceramic (SiO_x) and metallic oxide coatings in-line onto moving webs. If validated, coating with silica can produce the benefits of metallization without loss of transparency or the ability to microwave the film.

The following advantages are claimed:

Atmospheric plasma versus corona

- More durable treatment
- Superior surface energy and uniformity
- No backside treatment with low gauge films
- Treat porous materials with no pinholes
- Low level of ozone generation
- Treatment without oxygen
- Does not need a treater roll
- Cost is likely to be higher due to gas consumption

Compared with flame treatment

- Can be used to treat heat sensitive webs
- Chemistry of surface can be tailored per application specifications
- Less fire hazard
- No overtreatment

6.3 Ozone treatment

Over the years a variety of methods to accelerate melt oxidation have been evaluated and patented. Very few of these have proved to be practical. Ozone treatment of the polymer melt was one of the few ideas that were successfully exploited. This was encouraged by the development of powerful ozone generators for water treatment plants. This technology opened up the possibility of using such equipment on plastic converting machines. The precondition for the successful commercialization of ozone treatment for extrusion coating was the development of equipment to produce the ozone enriched air remote from the coating nip.

Figure 6.2 shows an ozone generating unit with remote applicator. The high frequency and high voltage is produced from the mains power in the generator. This is converted into a corona discharge, which oxidizes the incoming air to ozone. The ozonized air is then metered to the applicator, which is placed between the die and the coating nip.

The powerful oxidizing effect of ozone is due to the decomposition of the ozone (O_3) molecule into oxygen (O_2) and nascent oxygen (O). Nascent oxygen has a much shorter reaction initiation time than molecular oxygen and consequently accelerates the oxidation process. This effect is depicted in **Figure 6.3** in a dimensionless graph where the oxygen uptake of molten LDPE is plotted versus time.

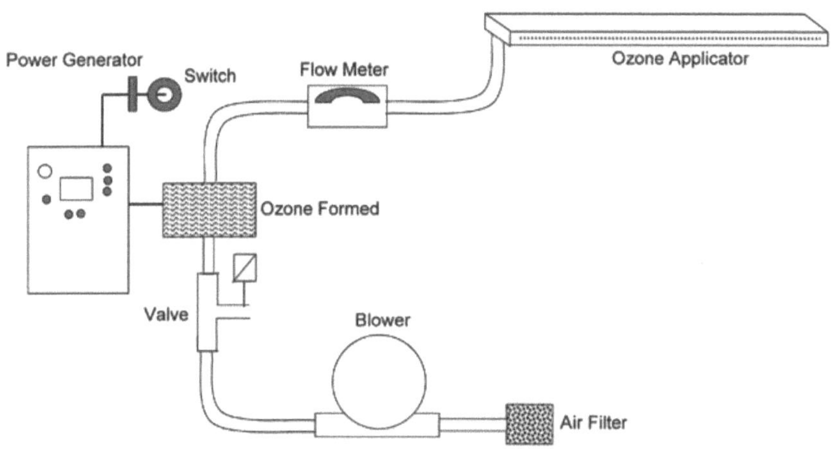

Figure 6.2

Ozone generator with remote applicator

Ozone is produced when **dry** air is passed between two electrical discharge elements one of which is maintained at 10 kV and the other grounded. The discharge elements can comprise a series of glass and metal plates or concentric glass and metal tubes with a gap across which the high voltage is applied. By careful adjustment of this narrow gap the high voltage discharge created appears to the naked eye as a violet glow.

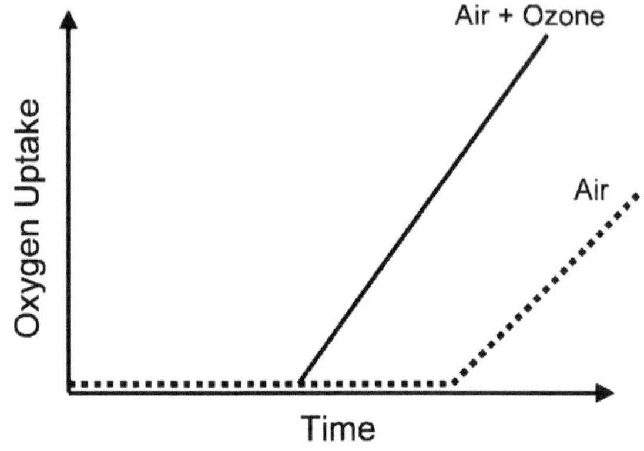

Figure 6.3

Oxygen uptake vs. time

Ozone itself is a hazardous substance and must be treated with care. Its odour is very readily detectable and has a 'threshold limit value for exposure' (TLV) of 0.1 ppm. Human sensitivity to ozone varies from 0.01-0.05 ppm. To reduce health risks the equipment must be designed to operate with negative pressure to eliminate ozone leaking into the working area. In sufficient concentration ozone can cause nausea and breathing difficulties. Ozone generators must be monitored and any excess gas must be flushed out of the area.

The position of the ozone applicator can be critical. A recommended starting position is shown in **Figure 6.4**. The distance from the melt is 4 cm (1.6″) and the height from the nip around 5-8 cm.

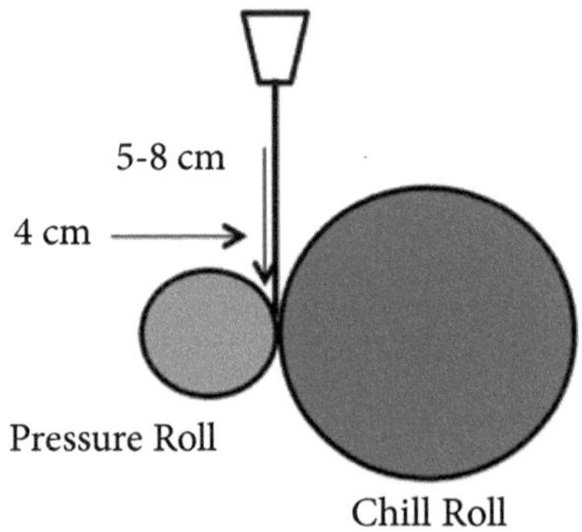

Figure 6.4

Position of ozone applicator in air gap

The applicator is a heat and corrosion resistant steel tube with usually 1 mm diameter holes drilled some 10 mm apart. These allow the gas stream to shower the polymer melt in the air gap. An air-knife with a 1 mm wide slot is an alternate design to the perforated tube.

The applicator should provide the following:

- Full width application on the melt
- Deckling facility adjustable with melt width
- Control velocity of ozone flow
- No disturbance of polymer melt

The plot in **Figure 6.5** shows the effect of ozone treatment on the adhesion of an LDPE (8 dg/min. melt index) coated onto aluminium foil at 315 °C. Coating weight was 40 g/m^2.

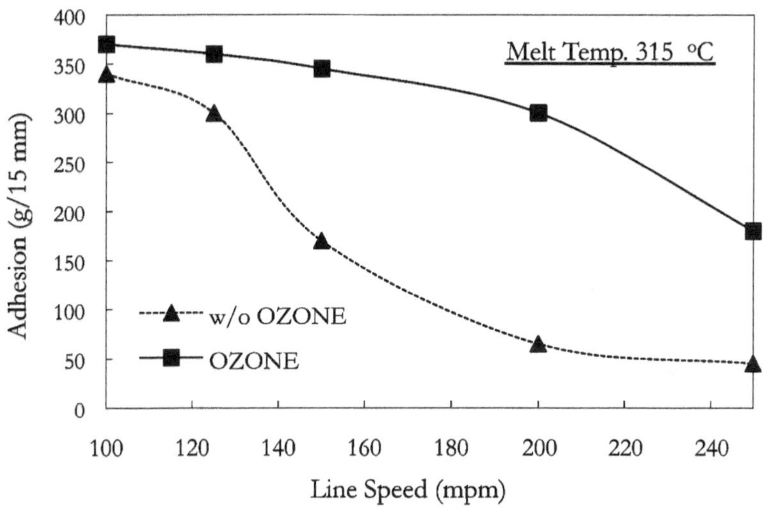

Figure 6.5

Adhesion to aluminium foil as a function of line speed

The graph clearly shows the effectiveness of ozone with increasing line speed. As melt residence time in the air gap is reduced the forced oxidation resulting from the presence of ozone enriched air maintains a higher adhesion level. Without ozone the decay starts at around 120 mpm. Similar data is shown in **Figure 6.6**, but this time extrusion temperature was varied. At a constant line speed of 100 mpm the LDPE coating has no adhesion to the aluminium foil **below** 310 °C.

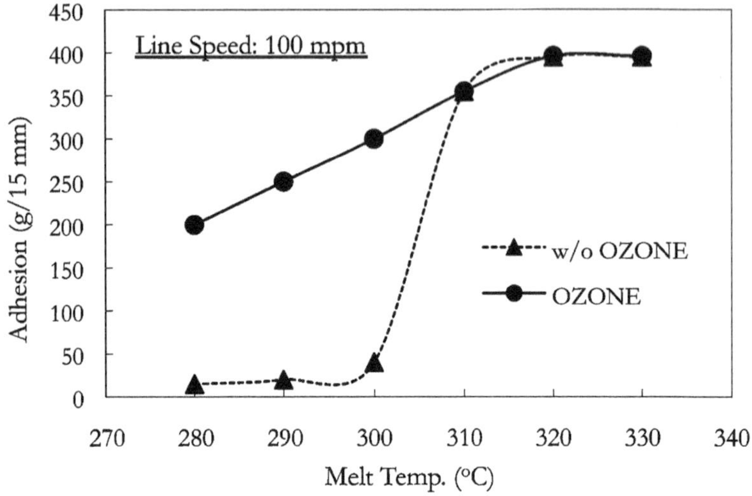

Figure 6.6

Adhesion to aluminium foil vs. temperature

With ozone, adhesion of 200 g/15 mm is achieved at 280 °C rising rapidly to reach a plateau at 320 °C where the curve meets the non-ozonized plot. These examples show that ozone treatment is an effective technique to either increase line speed or lower melt temperature to minimize LDPE odour for sensitive packaging applications. It is interesting to note that adhesion plateaus at 320 °C irrespective of the ozonizer. This confirms that once the melt surface is sufficiently oxidized no improvement in adhesion is possible on smooth surfaced substrates. Adhesion is limited by the relative strength of the adsorptive bonds generated between the oxidized polyethylene and the surface of the aluminium.

Note. *The use of ozone does not preclude the need for corona treatment to activate the surface of the substrate. The two techniques combined can allow the use of lower temperatures or higher line speeds on most substrates.*

A method to extrusion coat LDPE on polyamide films has been proposed by combining corona treatment with ozonization. The LDPE is ozone treated in the air gap and coated on the corona treated polyamide film. This intense oxidation generates enough dipole-dipole attractions that result in adhesion values acceptable for commercial exploitation. This method eliminates the need for a 3-layer coextrusion of the PE with polyamide using a costly tie-layer.

Ozonization equipment is relatively costly and requires constant maintenance to ensure optimum operation. Also the ozone applicator can be difficult to set-up in the limited space available between the die and the chill roll.

Studies have shown that ozonization is ineffective with EAA copolymers and works only with polyethylenes where the degree of oxidation is paramount[1]. *(This is an excellent paper and is recommended to the reader.)*

6.4 Flame treatment

Adhesion promotion by flame treatment is claimed to be superior to corona treatment under certain conditions. The techniques have some similarity since they both work by subjecting the substrate to an oxidizing environment, which increases surface tension. The advantages of flame treatment are as follows:

- Higher treatment levels (1600-1800 °C)
- Guaranteed no backside treatment
- Low decay rates
- No ozone formed
- No pinholing
- Surface cleansing of substrate
- Web widths up to 8 m

One of the disadvantages of corona treatment with very **wide** webs is the need to keep a perfect alignment of the electrodes and a precise gap setting across the width. This becomes more difficult as the web width increases. The electrodes must be free of debris and corrosion. In contrast the flame burner once correctly set up requires little maintenance.

Figure 6.7 shows the two zones of a burning flame. The inner zone is the area of incomplete combustion and is the coolest region of the flame. The hottest region of the flame is the outer zone. This is the area of complete combustion which contains the excited oxidizing species. The estimated temperature in this region is 1600-1800 °C. Addition of 5% by volume of oxygen to the air and hydrocarbon gas mixture can substantially increase the temperature of the oxidizing zone.

Flames are plasmas characterized by electron energies of 0.5 eV. Many excited species are present in a flame, e.g. free radicals, ions and electrons, etc.

Precise adjustment of the gas mixture to achieve maximum temperature in the oxidizing zone is the key to effective treatment. The distance of the substrate

1 Morris, B.A. and Suzuki, N., The Case Against Oxidation as the Primary Factor for Bonding Acid Copolymer to Foil, ANTEC 2001, SPE, **2001**, p. 25-35

from the flame is also important. The substrate must not come in contact with the reducing zone of the flame.

Figure 6.8 shows the basic layout of a flame treating installation. The equipment should be designed to allow precise control of angle and distance of the flame applicator from the substrate. The contact roll must be water cooled (30-50 °C) to remove excess heat.

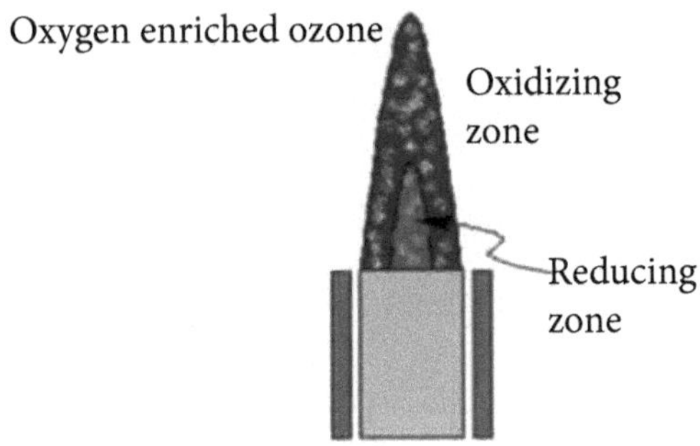

Figure 6.7

Schematic of flame structure

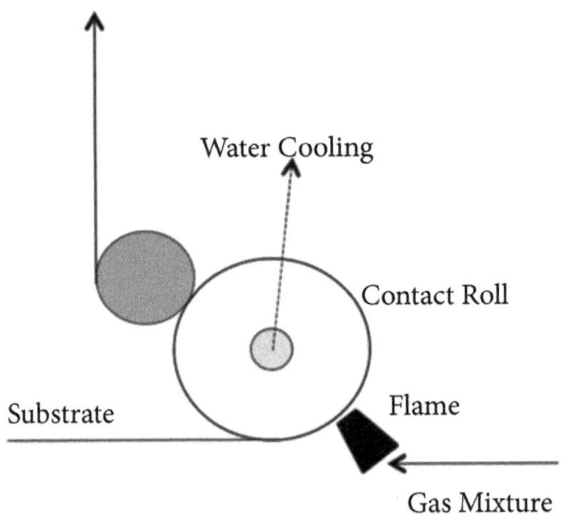

Figure 6.8

Schematic of flame burner installation

Figure 6.9 shows the maximum flame temperature or ionic activity as a function of oxygen in the gas/air premix. A critical element is the system to control the desired gas mixture ratio. The mixer is usually based on a Venturi device, which

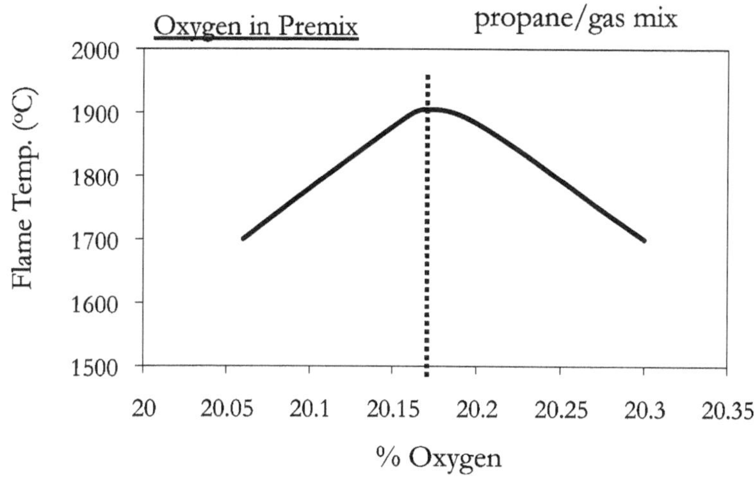

Figure 6.9

Flame treatment vs. oxygen in premix

feeds and doses air, hydrocarbon gas and additional oxygen if specified. A detector to measure and control the calorific value of the mixture must also be installed to stabilise the treatment process. The oxygen concentration can be controlled by building an oxygen analyzer into the system, which enables the user to adjust the mixture ratio to provide the desired surface treatment.

Highly sensitized safety features must be included to shut the gas off in emergencies and machine shutdowns.

The effectiveness of flame treatment will depend on the gas mixture and how the mixture is released. The recommended distance between the flame and the target is 7-8 cm. Paper/board coating requires more energy than plastic films.

Gas burners are generally used on large high-speed paper and paperboard coating machines. They are claimed to be more effective than corona treaters at very high surface area throughputs particularly with wide web widths. Flame treatment has the advantage of drying the paper surface and raising its temperature so that the substrate arrives warm and dry at the coating nip. Another advantage is the avoidance of unwanted **back treatment** of thin films (BOPP, OPET etc.), which can occur with corona treatment.

In **Figure 6.10** the adhesion of LDPE coated on flame and corona treated aluminium foil is compared. Samples were aged for one month. Ageing seems to improve adhesion levels. The difference between the two techniques is marginal in this example.

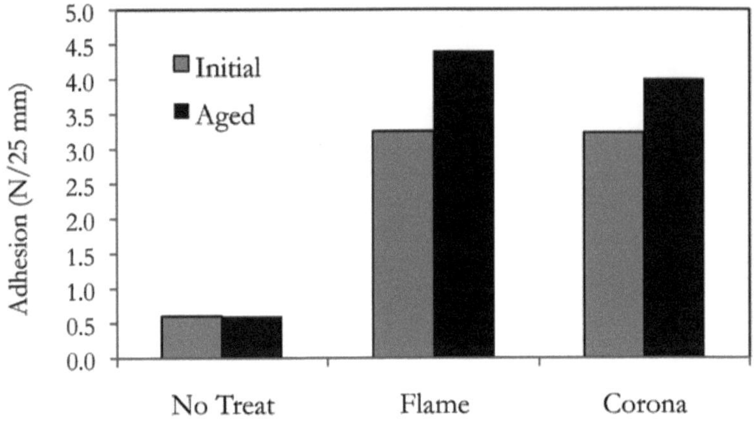

Figure 6.10

Adhesion of LDPE on aluminium foil (temp. 320 °C)

6.5 Chemical priming

There has been limited innovation in chemical primers for extrusion coating in recent years. There are a number of reasons for this:

- Film to film adhesive lamination has tended to be preferred over extrusion coatings for many multilayered products
- Copolymers such as EAA and ionomers have been developed, which achieve good adhesion to aluminium foil, one of the most frequently used substrates
- Advances have been made in coextrusion with feedblocks and effective adhesive tie-resins
- Air pollution controls have discouraged the development of new easy-to-apply solvent-based primer systems
- Machine line speed is restricted by the capacity of the drying oven

With the rapid development of corona and flame treatment techniques, the use of chemical primers has been restricted to non-paper-based films.

Some examples are:

- Polyamide/nylon films (oriented and non-oriented)
- Polyethylene terephthalate films (OPET)
- Regenerated cellulose films (RCF)

- Polypropylene films (BOPP)
- Metallized films
- Oxide (ceramic) coated films

In the case of aluminium foil coating, acid based copolymers (EAA) coextruded with LDPE are the preferred option.

Primers are divided into two basic types: aqueous and solvent based as shown in **Figure 6.11**. With solvent based primers explosion-proof equipment is mandatory for safety. The aqueous systems although obviously safer require more drying capacity to remove water.

The next level of subdivision is between the reactive and non-reactive or non-curing systems. The non-reactive formulations are usually preferred since they have no pot-life limitation and the maximum bond strength develops virtually off-machine. The curing systems require careful preparation and may take several days for the full bond to develop. Curing primers are used to ensure good heat and chemical resistance.

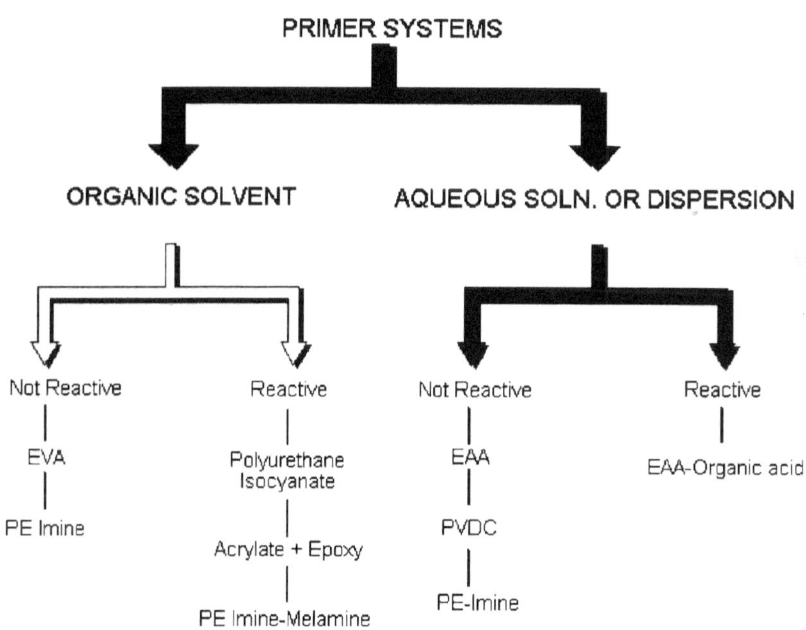

Figure 6.11

Chemical primer systems

Figure 6.12 shows the basic machine layout. The primer is applied to the substrate from a pan onto a roll and passed through a high velocity hot air drier. The dried surface is then fed into the extrusion coating nip. It should be remembered that the use of chemical primers does not alleviate the need for the melt to be sufficiently oxidized with enough **active sites** on its surface for the primer to interact. The primer enhances the reactivity with the melt at the interface and allows a greater tolerance in the extrusion temperatures that can be utilized.

Primers are usually applied on the substrate by smooth transfer roll techniques. These usually have low viscosity and are easily run on low cost applicators that work with aqueous primers. In some cases flexographic applicators are used using flexible rubber or plastic plates. These are widely used in printing on plastic films.

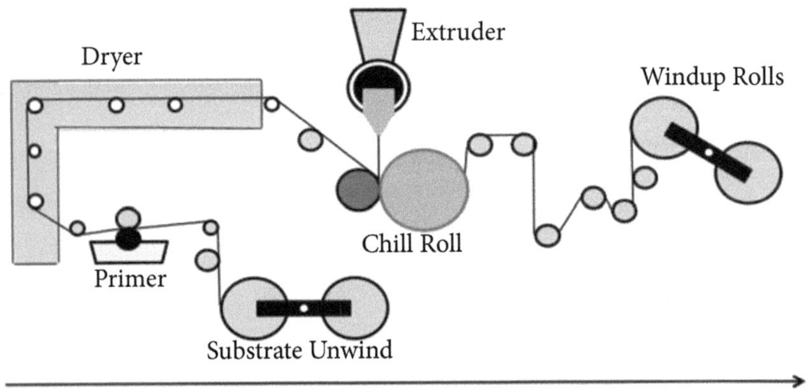

Figure 6.12

In-line chemical priming installation

6.5.1 Aqueous primers

Before the development of corona and flame treatment aqueous primers were originally used for application on paper and board. These are aqueous solutions based on organic imines and are applied to the substrate at coating weights of 0.02-0.08 g/m^2. A transfer roll applicator is adequate to apply the primer onto the high-speed moving substrate.

The PE-imines can be used with extrusion coated LDPE on:

- Paper/board
- Polyamide films

- Aluminium foil
- Metallized films
- Cellulose films (RCF)
- OPET films

The substrates should preferably be **corona** treated in-line. The imines also work well with EVAs, and other copolymers. Aqueous emulsions based on acrylic acid copolymers can also be used with LDPE coated onto aluminium foil, but these are more costly than the imine based formulations.

Table 6.3 lists some adhesion results obtained with an LDPE (8 dg/min. MI and density 0.916 g/cm^3) coated on a range of substrates at 60 mpm. Extrusion temperature was 295 °C. The data compares in-line corona treatment with an aqueous primer. The primer was applied with an 80 lines/cm gravure roll.

The PVDC coating on the MXXT/S cellophane film tended to lift off fairly easily and prevent an assessment of the effectiveness of the surface treatment. The results do however show the versatility of this particular water based primer. The greater effectiveness of corona treatment on the 24 μm BOPP film compared to the 12 μm sample is probably due to the improved affinity of the coextruded BOPP to both primer and LDPE. The copolymer coextruded with the PP may be influencing adhesion.

Table 6.3

Comparison of adhesion values

Substrate	Coating wt. (g/m^2)	No treatment (N/cm)	Corona (N/cm)	Primer (N/cm)
PET film	24	No adhesion	1.9	> TS
MXXT/S cellophane	24	No adhesion	>PVDC adhesion	> TS
DMS cellophane	24	0.3	> TS	> TS
PA-6 film	15	0.3	1.0	> TS
Al foil	24	0.5	-	> TS
OPP film	12	No adhesion	No adhesion	> TS
OPP (coex.) film	24	0.7	> TS	> TS
TS = Tensile strength of LDPE coating				

Figure 6.13 compares adhesion of corona treatment to chemical (aqueous) priming on aluminium foil at 310 °C melt temperature. The primer reaches adhesion values above 2.5 N/15 mm at a coating weight of 20 g/m^2.

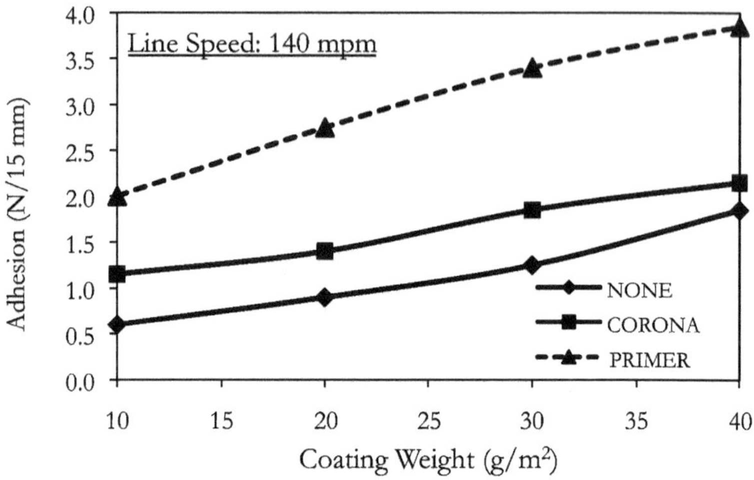

Figure 6.13

Effect of coating weight and treatment on adhesion

6.5.2 Solvent based primers

These are mainly 2-component curing systems and are designed to provide good adhesion and chemical resistance on plastic films and aluminium foil. Typical coating weights will average 0.3-0.5 g/m^2 dry weight. For exceptional product resistance the cured polyurethane primers applied at a coating weight of at least 1.0 g/m^2 are recommended. These high coating weights will increase cost. Direct gravure roll applicators are the most practical way of applying these primers. Drying is crucial in order to completely remove solvent residues and ensure a complete cure. It is essential that all the monomer is completely cured.

As a conclusion, if the extrusion coating application does not work with the simple aqueous primers and requires a more costly solvent based polyurethane system, it is recommended to consider film to film adhesive lamination as an alternative. Coextrusion coating with reactive copolymers is another option.

Useful further reading on adhesion promotion can be found in papers by Cheney et al.[2] and Greig and Jadon[3].

2 Cheney G., Markgraf D., Benson M., Statistical Analysis of the Effects of Ozone on Adhesion in the Extrusion Coating Process, TAPPI Polymers, Laminations and Coatings Conference, Toronto, TAPPI Press, **1997**, p. 649-655

3 Greig S., Jadon N., Sherman Treaters Ltd. Corona, Ozone and Flame Treaters for Extrusion Coating Lines, TAPPI European PLC Conference, TAPPI Press, **2001**

Chapter 7

The adhesive lamination process

The flexible packaging industry requires products that must meet a number of very demanding specifications. Especially when packaging perishable foods no single film will meet all criteria.

Film to film adhesive lamination offers many advantages compared to extrusion coating and coextrusion:

- Economical for short runs
- Reverse printing possible
- Excellent print registration
- Good gauge control of individual layers
- Films with excellent machineability
- Combines all roll-fed materials: PE films, oriented and metallized films, aluminium foil and paper
- Combines coextruded films with other supports

However, the technology has certain drawbacks. Some adhesives may require up to 10 days to complete the cure and solvents may also form VOCs (volatile organic compounds) which can impact on the environment.

The capital investment for a state-of-the-art extrusion coating plant is significantly higher than a laminator. However, if the investment for the extrusion plant to produce the lamination film is included then the investment costs begin to converge. In terms of operating costs lamination adhesives are higher priced than polyethylene and the copolymers used in extrusion coating. Minimizing adhesive coating weight is critical in cost control. In adhesive lamination energy is required for drying and pollution free solvent recovery is mandatory. On balance the decision favouring one process will generally depend on the flexibility required in terms of material choices and production run lengths. Extrusion coating is not viable for short runs. Extrusion coating machines are typically 1.2-2.5 m in width whereas laminators are usually 75-80 cm wide.

Coextrusion is another option, but for demanding high performance applications specifying bi-oriented and metallized films and foil, coextrusion is inadequate.

For perishable food packaging where high barrier protection is paramount the cost of coextruded films with EVOH or other barrier polymer can be high. In such cases a minimum 5-layer machine will be required to include the costly tie resin layers. If the blown film process is used the investments in multilayered tubular dies with the required number of extruders will also be high. The setting up and maintenance of these dies is difficult and time consuming. Flat die chill roll processes with combining feedblocks are more practical and easier to operate.

In adhesive lamination a number of considerations must be taken into account in the selection of the bonding system and equipment design. These are listed below:

- Adhesive curing time and temperature
- Mixing ratio of components
- Solution viscosity
- Drying velocity
- Initial bond strength (green strength)
- Nip pressure and temperature

Figure 7.1 shows the layout of a basic film lamination machine. The adhesive solution is spread from a gravure roll applicator onto the film (primary web) or substrate. The gravure roll is usually a steel roll or cylinder engraved with cells that pick up the liquid, which is applied on the web and passed through a nip. The application roll will usually have a surface finish of 40-70 lines/cm (100-175 lines/inch). Typical solvents are ethyl acetate, methyl ethyl ketone (MEK) and alcohols. The adhesive should have a surface tension at least 10 mN/m **lower** than the surface being coated.

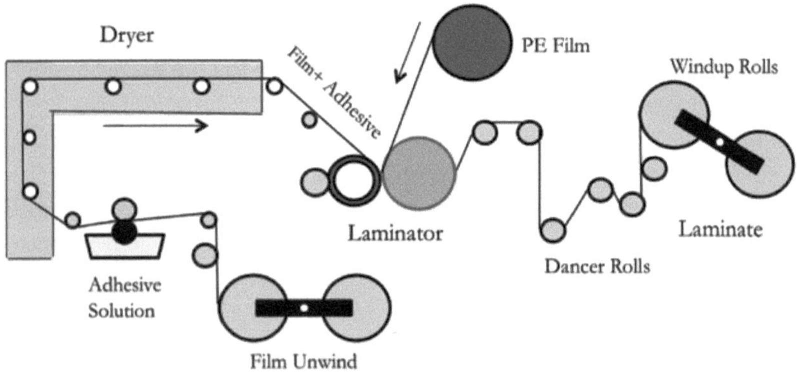

Figure 7.1

Adhesive laminator

The dry coating weight, which can be as low as 1 g/m^2, will depend on the applicator configuration and the viscosity of the adhesive solution. Drying is carried out in a tunnel usually with high velocity hot air. The dried film is then laminated to the polyethylene film in a heated nip (40-70 °C) as shown in **Figure 7.1**. A steel back-up roll to prevent deflection is normally included. To avoid film curling, the two films entering the nip must be at equal tension. DC drives capable of accurate tension control must be used. It is important that the adhesive layer is not rubbed by contact with rolls or other surfaces prior to entering the laminating section. Some typical adhesive applicator configurations are shown in **Figure 7.2**.

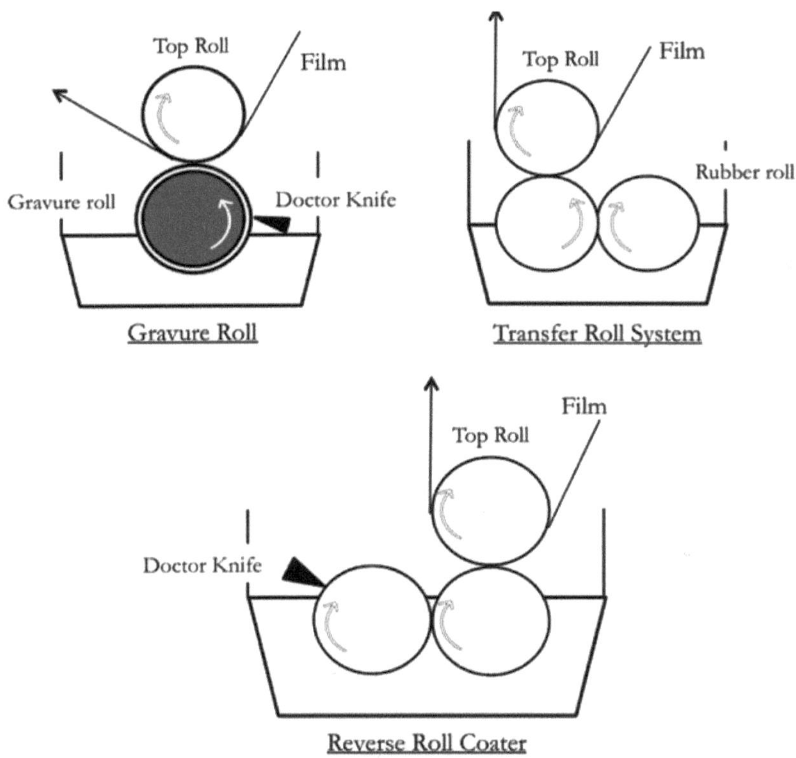

Figure 7.2

Typical adhesive applicators

Since the adhesive will generally have an initial low tack, it is important that tension control is precise to avoid any delamination between the laminating nip and the winder. The line should be equipped with dancer rolls, either pneumatically or electronically operated, to equalize web tension in both the machine and cross directions and avoid curling and in the worst case the formation of creases and wrinkles. Ageing for a week is normally required to complete the cure of isocyanate based adhesives.

To avoid roll deflection the roll width to diameter ratio as shown below is typical. Nip pressure should be in the range of 10-50 kg/cm width. The nip rolls should be perfectly parallel with uniform pressure distribution across their width.

Some roll specifications:

Top roller
Durometer hardness Shore A: 35-40
Sleeve type: neoprene, silicone or polyurethane
Bottom roller
Nickel or chrome plated steel or hard rubber
Typical roll width: 75-80 cm
Roll diameter: 15-20 cm

Table 7.1 shows some of the problems that can occur during the lamination of two films, and possible solutions.

Table 7.1

Lamination problems and solutions

Defect	Adjustment required
Wrinkles, poor bond, web curl	Tension control
Internal laminate stresses causing web curl	In-out feed tension control
Weak bond, adhesive ooze, wrinkles	Nip adjustment, tension control
Air bubbles, wrinkles	Contact point: poorly aligned rolls

Figure 7.3 shows possible defects that can occur with poorly machined or maladjusted nip rolls. Adhesion will be uneven across the web from canted, concave or convex shaped rolls. Other defects are curling, bubbles and wrinkle formation. Telescoping of the laminated rolls can occur with low web tension, excess slip additive and low green strength (initial adhesion) between layers.

At the nip the rolls must perfectly flat and parallel with uniform pressure across the width.

Nip pressures can be controlled manually or pneumatically with air cylinders. There should be a single control for rapid and uniform adjustment of the actuators. Non-uniform pressure will also lead to poor web steering in addition to the above defects. The PE film must be brought in contact with the adhesive layer at the nip contact point and not earlier otherwise air entrapment can occur.

Figure 7.4 shows schematically a lamination combining reverse printed OPET with LDPE film. It is important to ensure that there are no interactions between the adhesive, the printing inks and additives in the film.

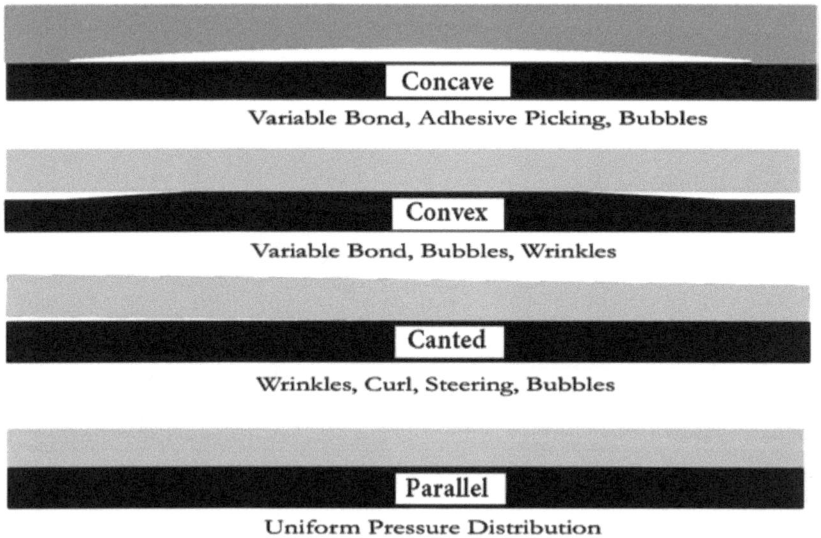

Figure 7.3

Faults at interface of applicator rolls

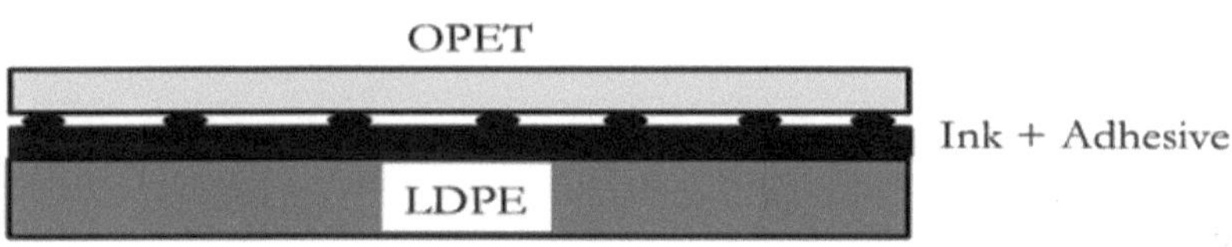

Figure 7.4

Typical laminated structure

7.1 Adhesives

The individual reactants of a multi-component adhesive are not adhesive by nature. These react with each other after mixing and develop strong bonds only on curing. A crosslinking reaction is usually propagated by heat. The multi-component resins can be either solvent-based or solventless. The solvents in the formulation act as diluents for the resins, which facilitates wetting and spreading on the web. The solvent must be completely devolatilized during the curing process.

The adhesives used in high performance film lamination are available in the following categories:

- Solvent-based 1 component
- Solvent-based 2 components
- Solventless 1 component
- Solventless 2 components
- Aqueous/water borne

A snapshot comparison of the above systems is shown in **Table 7.2**.

Table 7.2

Subjective comparison of basic adhesive systems

Solvent-based	Aqueous	Solventless
Urethane based	Urethane based	Urethane based
Solvent emissions	No organic solvent	No emissions
Medium adhesive cost	Medium adhesive cost	Low adhesive cost
High energy needs	High energy needs	Low energy needs
Medium line speed	Medium line speed	High line speed
Low viscosity	Low viscosity	Medium viscosity
Curing during lamination	Curing after lamination	Curing after lamination

The most commonly used adhesives are those based on polyurethane chemistry. A urethane is formed by reacting an isocyanate with a hydroxyl containing reagent (–OH) such as a polyol.

R-NCO + HO-R′ → R-NH-CO-O-R′
Isocyanate + hydroxyl **urethane**

A typical polyurethane is synthesized by reacting a di-isocyanate, typically toluene di-isocyanate (TDI) or 4,4-diphenylmethane di-isocyanate (MDI), with a diol as follows:

OCN-R-NCO + HO-R′-OH → -(-CO-NH-R-NH-CO-O-R′-O-)
Diol **linear polyurethane**

The chain is extended by reacting the di-isocyanate with hydroxyl terminated polyesters or polyethers. These are available as single or 2-component adhesives. The polyether-urethanes are usually more heat resistant than their polyester counterparts.

1-component adhesives

These adhesives are based on low molecular weight pre-polymers terminated with isocyanate groups. The isocyanate groups are free to react with atmospheric moisture or with a mist spray at the laminator nip. Carbon dioxide is released from each mole of isocyanate.

NCO---------NCO + H_2O → -----------NH_2 + **CO_2↑** + NCO---------NCO
prepolymer **amine** **prepolymer**
↓
---------NH-CO-NH---------
polyurea

The release of carbon dioxide may cause blistering with non-porous films.

2-component adhesives

2-component adhesives are produced by mixing a pre-polymer with free isocyanate groups with a resin containing free hydroxyl groups (polyol). The 2 components will polymerize to form an adhesive film. Excess isocyanate will react with moisture in the atmosphere and/or substrate.

NCO---------NCO + OH---------OH → --------NH-CO-O--------
prepolymer **polyol** **high polymer**

The isocyanate component will usually have a viscosity of 1.6-3.0 Pa.s and the resin component a viscosity of 5.0-15 Pa.s.

In order that the adhesive be effective, a threshold molecular weight must be achieved. With low reactivity polyolefins films high molecular weights are required to achieve useful properties. Typical molecular weights of urethane adhesive systems are shown in **Table 7.3**.

Table 7.3

Molecular weights of urethane adhesives

Adhesive type	M_w	M_n
Solvent-based	12,000	6,000
1 part solventless	9,000	5,000
2 part solventless	6,000	4,000
Aqueous	72,000	40,000

The aqueous adhesives are dispersions and can, therefore, be polymerized to high molecular weights, as their viscosity in water will be low enough to be applied from a roller coater. In contrast the solventless systems are prepolymerized to controlled molecular weights in order to keep their viscosity sufficiently tractable for application. There is no solvent present to act as a diluent and wetting agent.

The components must be well mixed and in the correct proportions and evenly spread on the film surface. The ratio of isocyanate to hydroxyl and the laminating and storage conditions will influence the quality of the bond. The lower the viscosity the easier it is to spread the adhesive evenly and achieve a uniform dry weight on the surface. The adhesive bond is made between active radicals on the film's surface and the reactive component from the adhesive. In the case of polyethylene, a corona treated surface with a surface tension of 42-45 mN/m is normally required for interaction with the adhesive.

The key requirements of laminating adhesives are:

- Non toxic
- Good green strength (initial adhesion)
- Fast solvent release
- Film-forming at low coating weight
- Good wetting
- Long pot life after mixing
- Absence of volatile by-products
- Product resistance
- Final bond strength

In the finished laminate the adhesive must fulfil the following criteria:

- Permanent bonding
- Heat stability
- Elasticity retention
- Colour stability
- Compliance with food packaging regulations

With the drive towards reduced environmental pollution from solvents, attention has focused on solventless formulations. Because of their higher viscosity these systems are more difficult to apply uniformly at the desired coating weights of

1-2 g/m^2. Solvents will also wet the film surface and wash away contaminants thus enhancing interface contact. This added benefit is not applicable with solventless formulations.

7.2 Solvent-free adhesives

Environmental pressures are driving converters to reduce their use of organic solvents. Adhesive suppliers have reacted by offering solventless laminating adhesives that cover a wide range of films and foils with good product resistance. These are however, still somewhat poorer in performance compared to the best solvent based formulations. One advantage of solventless systems is that they do not require a drying oven, which saves space and energy. Application weights are 1.5-4.5 g/m^2, which will increase material cost.

As outlined in the preceding section solvent based adhesives are prepared in two steps. A urethane pre-polymer is first produced and then diluted with a solvent prior to the molecular weight increasing to an unmanageable level by chain propagation. The diluent keeps the overall viscosity low for easy application on the substrate. The molecular weight is controlled prior to application and then cured in the final laminate.

In the solventless system the reaction in the first step is stopped earlier. The prepolymer has a lower molecular weight and low enough viscosity to be applied on a gravure or similar coater. Curing takes place at ambient temperature in the laminate. Since the solventless system is applied at a lower molecular weight more polymerization takes place after application in the laminator.

Problems may arise with 1-component isocyanate terminated systems since these are cured with moisture. A moisture spray is usually installed on the machine. The risk here is that reaction may occur with retained alcohol in printing inks rather than moisture, leading to an uneven cure. However, with care these solventless 1-component adhesives can be used with a variety of substrates providing these are lightly printed. Generally polyether based adhesives are preferred for film and paper substrates and work well with highly printed surfaces. For aluminium foil laminations only polyester systems should be used.

In a 2-component formulation where curing takes place between the isocyanate and the hydroxyls in the curing agent this problem is less pronounced. Solventless 2-component adhesives are available in two forms:

- Low viscosity applied at ambient temperature
- High viscosity applied at 40-70 °C

High viscosity adhesives have the advantage of developing higher initial bond strengths. The low viscosity systems are used with heat sensitive films, but because of their low initial molecular weight they will have lower initial bond (green) strength.

To optimize surface wetting and have minimal effect on the slip additives present in the films, solventless formulations based on combinations of polyether and polyester groups are used. These are claimed to have a minimal effect on COF and are recommended for high-speed form, fill and seal applications. Although the polyether systems wet better and develop strong bonds they tend to absorb migratory molecules such as slip additives and waxes usually present in PE. The polyesters on the other hand do not absorb slip additives and, therefore, have a minimum effect on the laminate's coefficient of friction (COF). Combining the two functional groups into one system is one approach to optimizing the performance of the laminate.

Figure 7.5 is a schematic layout of a solventless adhesive applicator. The primary web is coated with the adhesive and laminated to the polyethylene film (secondary web) in the laminator nip as shown. No drying oven is required. The nip rolls are heated to 40-70 °C.

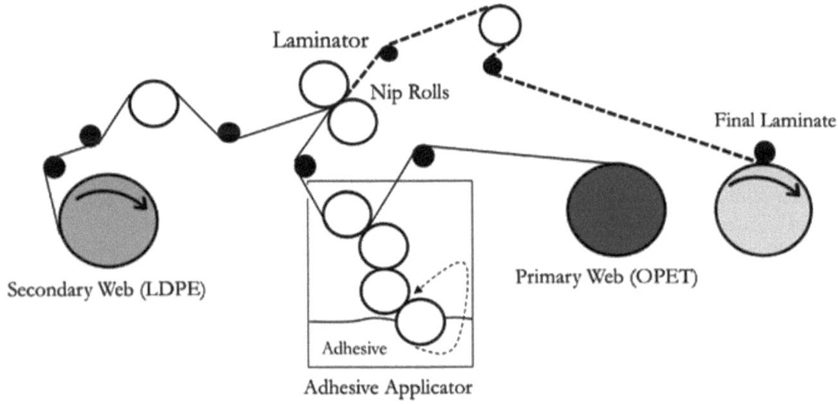

Figure 7.5

Schematic of solventless adhesive laminator

In **Table 7.4** the dependence of the laminate's coefficient of friction on slip agent (erucamide) concentration in the PE film is shown. The aluminium foil/LDPE laminates were produced with two 2-component solventless adhesives. The results show that adhesive #2 develops a too high COF with erucamide levels below 500 ppm after 1 month ageing. Adhesive #1 appears not to be sensitive to the slip agent concentration.

In **Table 7.5** some adhesion and heat seal strength data for a number of laminated structures are shown. A solventless 2-component polyether-polyester adhesive was used in all cases.

Table 7.4

COF and slip agent concentration

Al foil/LDPE	1 week	1 month
	150 ppm erucamide	
Adhesive #1	0.21	0.21
Adhesive #2	0.21	0.42
	300 ppm erucamide	
Adhesive #1	0.2	0.19
Adhesive #2	0.2	0.32
	500 ppm erucamide	
Adhesive #1	0.21	0.19
Adhesive #2	0.19	0.18

Table 7.5

Adhesion and seal strength of some typical laminates

Laminate	Bond strength after 8 days (N/15 mm)		Seal strength (N/15 mm)
	Printed	Unprinted	
OPA 15 µm/EVA 50 µm		4.5	34
PA 30 µm/EVA 50 µm		4.7	35
OPA 19 µm/EVA 50 µm	3.0	Inseparable	33
OPET 12 µm/LLDPE 50 µm	3.1	Inseparable	45
OPET 12 µm/Alu foil 12 µm	2.5	2.9	-
Met. OPP 19 µm/OPP 19 µm	1.9	2.6	-
Met. OPET 12 µm/LLDPE 60 µm	-	3.2	41

Figure 7.6 shows some heat seal values for LDPE and EVA films laminated with a 2-component solventless adhesive to 50 µm polyamide films. The polyolefin films were 100 µm in thickness and contain no slip additives. Lamination was carried out at 40 °C. The adhesive was an MDI prepolymer with a polyol curing agent (ratio 100 : 35). The EVAs develop strong seals at 120 °C whereas the LDPE needs a temperature above 140 °C. This example shows the importance of film selection to maximize heat seal strength at the lowest temperature.

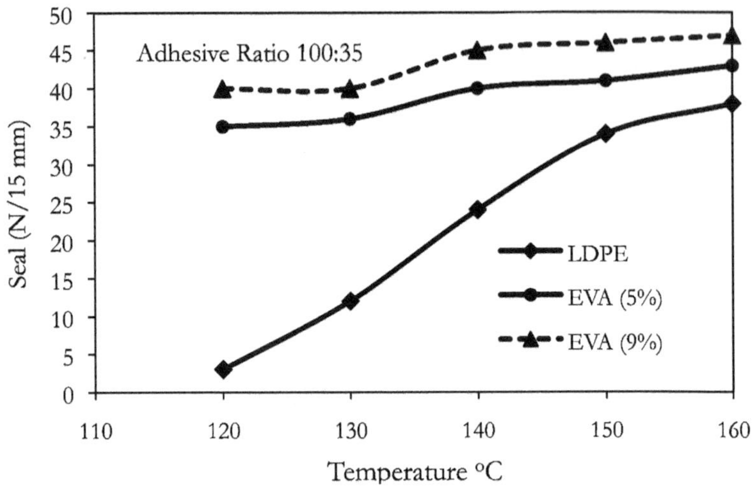

Figure 7.6

Seal strength to polyamide films

7.3 Product resistance

Laminates are heterogeneous structures, which embody different films, adhesives and inks. All or some components may contain migratory products such as slip agents, waxes, residual solvents and plasticizers. These materials may produce odours or off-taste, and affect the seal or bond strength and machineability on the packaging lines. It is therefore important that all materials are carefully selected to avoid any long term defects appearing in the finished package.

For food packaging the legislation in any country is very precise but essentially states that no component is allowed to migrate into the food so packaging will not impair its composition or its organoleptic properties.

Figure 7.7 shows a combination printing and lamination line to produce a typical BOPP based snack food structure. Printing ink formulations should be free of aromatic solvents (toluene) and lead chromate pigments. Other undesirable materials are chlorine-containing binders, phthalate and diarylide pigments.

The adhesive must completely wet the film surfaces including the printed areas. The inks must be completely dried and well adhered to the surfaces. There should be no interaction between the adhesive and the ink formulation. As discussed elsewhere, the polyethylene film must be correctly formulated with slip and antiblock additives. Surface treatments such as corona or flame treatment must be optimized for all film surfaces. Flame treatment is preferred by some laminators

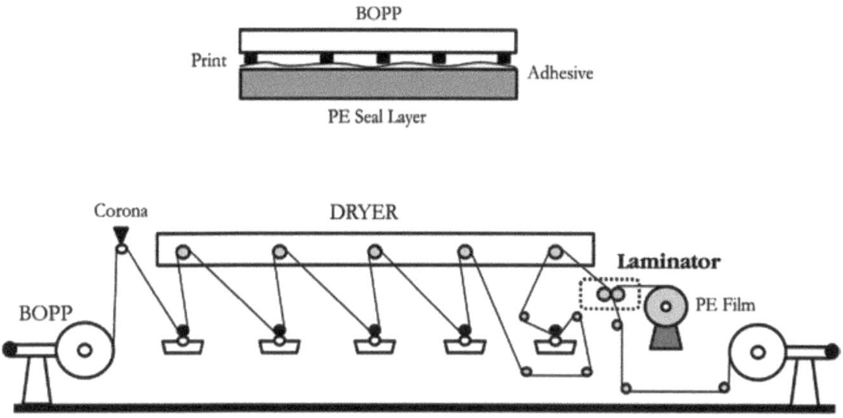

Figure 7.7

In-line printing and lamination

to avoid the risk of back treatment and pinhole formation on very thin films. More recently there has been an interest in atmospheric plasma treatment.

Excess slip additives may weaken the bond and seal strength. The reverse will lead to snagging on form, fill and seal machines. Acid or oily foods may attack the adhesive causing delamination. Therefore product resistance data must be generated for the proposed application. Some alkaline and acid foods may attack aluminium foil causing failure of the laminate. In these circumstances alternate barrier materials, such as PVDC, EVOH or ceramic coated OPET may be preferred.

Heat resistance of the structure must be verified if the packed product is to be pasteurized or sterilized.

Product resistance data for a 2-component solventless adhesive are shown in **Table 7.6**. This adhesive can be used with a wide range of film and foil combinations and can be classed as a medium performance adhesive comparable to solvent based systems. The adhesive was used to produce the following structure:

OPET 12 μm
Al foil 8 μm
LLDPE 60 μm

These results show poor product resistance to acids. Aluminium foil is attacked by acids, which explains the poor results in acidic media. Adhesion values in contact with motor oil were excellent and marginal to alcohols. The alcohols were tested at high concentration levels. In most toiletry products alcohol content will be lower.

In **Table 7.7** similar data is generated, but this time with an OPET (12 μm)/ LDPE (70 μm) laminate. The LDPE contains 350 ppm erucamide. Resistance to acids is now much improved by eliminating the aluminium foil. Resistance to alcohols is poor.

Table 7.6

Product resistance of laminate with solventless adhesive

Product	Bond strength (N/15 mm width @ 100 mm/min Xhead speed)	
	1 week @ 45 °C	1 month @ 45°C
Air	7.0-11.0	10
Water	4.8-6.3	6.5-8.0
Curry powder	2.0-2.5	1.5-2.5
Paprika powder	3.0-5.0	2.5-3.5
Shampoo	4.0-5.5	4.0-5.5
Liquid detergent	2.3-3.0	2.3-3.0
Mayonnaise	2.5-3.3	2.3-3.0
Fruit drink	4.0-5.5	3.5-4.8
Acetic acid (1%)	0.5	0.4
Isopropanol (90%)	0.9-1.3	1.2-1.5
Ethanol (99%)	0.8-1.0	1.0-1.1
Motor oil	5.8-8.0	5.7-8.0

Table 7.7

Product resistance of laminate with solventless adhesive (no Al foil layer)

Product	Bond strength (N/15 mm width @ 100 mm/min Xhead speed)	
	1 Week @ 45 °C	1 Month @ 45 °C
Air	N. S.	N. S.
Water	N. S.	N. S.
Curry powder	2.5 tear	N. S.
Paprika powder	2.1 tear	2.8 tear
Shampoo	N. S.	N. S.
Liquid detergent	N. S.	N. S.
Mayonnaise	N. S.	N. S.
Fruit drink	N. S.	N. S.
Acetic acid (1%)	N. S.	N. S.
Isopropanol (90%)	0.9-1.5	0.6-1.5
Ethanol (99%)	0.1	0.1
Motor oil	N. S.	N. S.
N. S. = not separable		

It is important that solventless adhesives have the minimum residues of free isocyanate or low molecular weight fractions in order to both meet food compliance legislation and minimize the risk of surface contamination, which might compromise seal integrity.

7.4 The anti-sealing effect in laminates

It is important that the adhesive is completely cured and none of the monomers are free to migrate to the polyethylene surface. This will compromise the heat seal strength of the film. This anti-sealing effect was first reported in the late 1970s at a time when laminators were trying to replace toluene di-isocyanate (TDI) with 4,4-diphenylmethane di-isocyanate (MDI) in adhesive formulations. This development was spurred by the tightening of legislation regarding the use of suspected carcinogens in the work area. MDI was preferred because of its lower vapour pressure at ambient temperatures. However, adhesive systems based on MDI were believed to contain a higher level of extractables increasing the risk of migration of unwanted substances at the film surface leading to undesirable after effects, such as resistance to heat sealing (anti-sealing).

Anti-sealing manifests itself as loss in seal strength of the laminate at the normally accepted seal temperature, time and pressure. The phenomenon occurs sporadically making it difficult to isolate its origins. Analysis of the non-sealing film surface has shown the presence of polyurea. Sealability is usually restored after washing the surface with a solvent, such as isopropanol. The presence of MDI monomer in the adhesive was believed to be the culprit. Un-reacted MDI migrates to the surface of the polyethylene seal layer, reacts with moisture and forms the amine by splitting off carbon dioxide (CO_2). The amine then reacts with residual di-isocyanate to form urea and polyurea.

The generic reaction is shown below:

$$\underset{\textbf{di-isocyanate}}{NCO-R-NCO} + 2H_2O \longrightarrow \underset{\textbf{amine}}{NH_2-R-NH_2} + \uparrow 2CO_2 \xrightarrow{(+\,NCO\text{-}R\text{-}NCO)} \underset{\textbf{polyurea}}{(R-NH-CO-NH-R)_n}$$

The molecular weight of the polyurea at the surface will determine the resistance to sealing at a given temperature.

Figure 7.8 compares the effect on seal strength of the MDI with and without the polyol curing agent. The drop in seal strength is catastrophic.

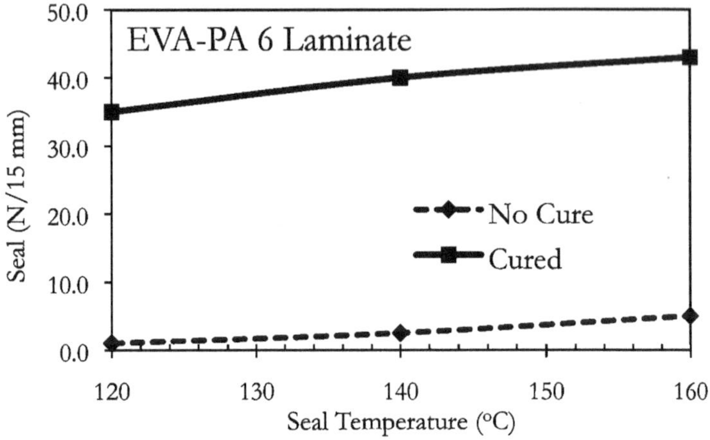

Figure 7.8

Effect of curing agent on seal strength

Adhesive suppliers are continually developing faster curing formulations that do not retain any free isocyanate that may contaminate the surface of the laminate.

It must also be borne in mind that slip (erucamide) concentration may also have an effect on seal strength. **Figure 7.9** shows the effect of slip agent concentration on the seal strength of an EVA (5%) film laminated with an MDI solventless adhesive to a 50 μm polyamide film. The EVA film was 100 μm in thickness.

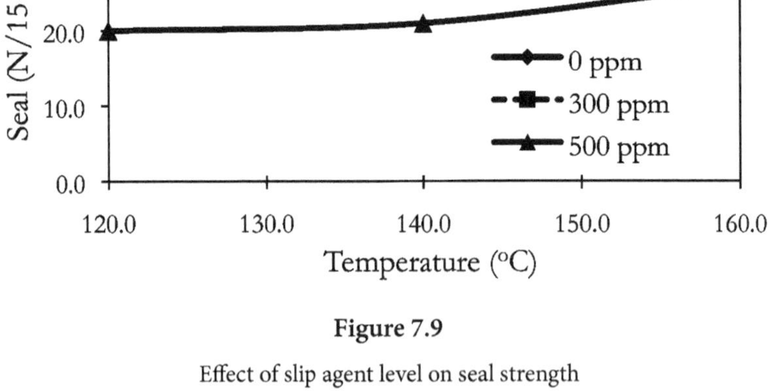

Figure 7.9

Effect of slip agent level on seal strength

This data shows that the presence of slip at concentrations above 300 ppm may also create an anti-sealing effect at this film thickness. The lamination film used in these experiments was 100 µm. If lower thickness films are used a slip concentration above 300 ppm may be required. The thinner the film the less slip agent will be available to migrate to its surface.

The anti-sealing phenomenon is clearly quite complex.

Chapter 8

Polyethylene lamination films

Lamination film specifications vary according to the type of goods packed, weight per pack, machinery used, sealing jaws, transportation and shelf stability. This is an important high value added market for polyethylene films. The PE film is used primarily as a heat seal layer laminated with a polyurethane adhesive or other adhesive to a support(s) such as:

- OPET film
- BOPP film
- Metallized film
- SiOx (oxide) coated films
- Polyamide including OPA (oriented) films
- Cellulose films (RCF)
- Paper

There are applications where the polyethylene film is extrusion laminated with LDPE to substrates, such as paper and aluminium foil combinations, where the polyethylene acts as the adhesive layer. Extrusion coating is the recommended process for these products.

8.1 Tubular film production

Figure 8.1 shows a typical layout for a tubular extrusion process to produce polyethylene film. The extruders are usually fitted with barrier screws (see section 3.4). Melt temperature is set in the range of 220-240 °C. Barrels should be designed with air cooling to avoid overheating particularly with LLDPE at high outputs. Grooved feed bushings are recommended by most manufacturers for polyethylene extrusion.

The melt is extruded into a tube, which is expanded with air to a blow-up ratio of approximately 2 : 1 and air cooled to 30-40 °C as it travels past the cooling ring placed above the die exit. The cooled bubble enters the collapsing frame and is fed to the nip roll section. The collapsing frame must be designed to avoid drag and chatter damage. Past the nip the layflat (double layer) is slit and separated into two

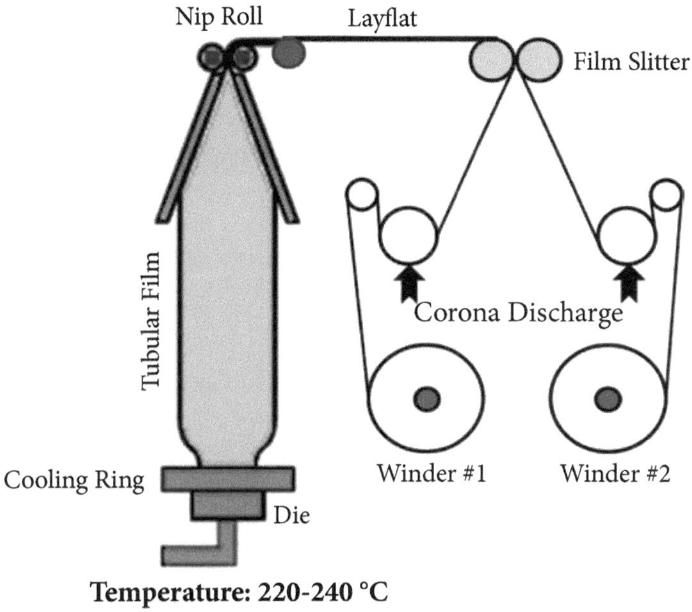

Figure 8.1

Tubular film plant for polyethylene

layers. These are surface treated by corona discharge or other technique and fed to separate winders. Typical outputs are 2.3 kg/hr./cm of die, e.g. 250 kg/hr. for a 350 mm diameter die. The limited cooling capability of air restricts the output of the process. Chilled (10 °C) air is used to increase output. Some machines include internal bubble cooling (IBC), which can further increase productivity.

The air ring must be carefully adjusted to control the frost line height and to evenly cool and feed the tube into the collapsing frame. Film thickness is determined by the extruder output, the haul-off rate and blow-up ratio. Extruder sizes are usually in the range of 90-114 mm (3½-4½″) diameter. Coextrusion lines will normally include smaller (40-60 mm diameter) satellite extruders.

The principle underlying internal bubble cooling (IBC) is to apply a flow of cold air both inside and outside of the bubble. By this means the bubble is more efficiently cooled and stabilized by the opposing air currents. Leading manufacturers of IBC systems claim an increase in output of more than 50%. When used with LLDPE the IBC unit is claimed to achieve outputs of 3-4 kg/hr./cm of die compared to 2.3 kg/hr./cm with a conventional dual orifice cooling ring.

Machine output can be increased by cooling the air flowing into the cooling ring. A computerized simulation was carried out to evaluate the effects of two air temperatures on output. The die size was 250 mm diameter delivering 200 kg/hr. of 75 μm film.

Table 8.1 shows that at 200 kg/hr. of LDPE output the film cools to 40 °C at a height of 258 cm from the die face with an air temperature of 20 °C. If the cooling air is chilled to 10 °C, output is increased to 245 kg/hr. with the film cooling to 40 °C at very nearly the same distance from the die face. This simulation shows an output increase of 22.5% illustrating the benefits of using refrigerated air.

Table 8.1

Effect of air temperature on output

Cooling ring air temperature	20 °C	10 °C
Film thickness	75 μm	75 μm
Melt temperature	220 °C	220 °C
Frost line	750 mm	750 mm
Line speed	31 mpm	38 mpm
Film height @ 40 °C	258 cm	255 cm
LDPE mass flow rate	200 kg/hr.	245 kg/hr.

Another simulation was carried out comparing the IBC with a conventional cooling ring. The details are in **Table 8.2**. With a standard cooling ring the LDPE film temperature at 1 m distance from the die face is 98 °C. With an IBC programmed into the experiment the output from the die was increased from 200 kg/hr. to 250 kg/hr. with the same film temperature also at a distance of 1 m from the die face. This result represents a 25% increase in output. Air temperature was fixed at 20 °C in both cases. This suggests that claims of 50% increases in output may be optimistic unless refrigerated air is used with the IBC.

Table 8.2

Effect of internal bubble cooling on output

Cooling	Normal	IBC
Cooling ring air temperature	20 °C	20 °C (in/out)
Film thickness	30 μm	30 μm
Melt temperature	200 °C	200 °C
Frost line	750 mm	750 mm
Line speed	65 mpm	82 mpm
Film temperature @ 1 m	98 °C	98 °C
LDPE mass flow rate	200 kg/hr.	250 kg/hr.

These virtual experiments indicate that the use of chilled air is almost as effective as the use of an IBC system. Combining the two when possible is recommended.

The films must be surface treated to 42-45 mN/m and be pinhole free. Coextrusion eliminates the risk of pinhole formation. Thickness variation should be kept below ±8% of the target value. For some applications it may be necessary to reduce the variation to <5%. During the lamination process it is essential that the film is perfectly flat as it feeds into the laminator. These tight gauge requirements can only be met with precision designed dies. These are usually based on spiral mandrels to evenly distribute the melt round the annular channel and avoid weld lines. Oscillating dies or rotating haul-offs are necessary to ensure flat rolls free of gauge bands (random distribution of thickness variations across the roll width). The collapsing and winding systems must deliver rolls wound at uniform tension and completely free of wrinkles and creases. Centre or gap winders are used. It may also be necessary to edge trim (~10 mm) the layflat so that the finished rolls have perfectly aligned edges. If this is done the trim should be recycled in-line to minimize scrap losses. Devices that collect, compact and chop up the trim should be included in the machine design. The chopped up trim is recycled via a screw transport feeder back into the extruder. Most lines used today for high performance lamination films incorporate coextrusion: the trim is best recycled in the core layer of an ABA or ABC structure. Some food packaging legislation may proscribe the use of recycled or scrap material.

The slip agent concentration must be accurately controlled. Erucamide is preferred over oleamide. Too much slip will weaken adhesion causing delamination. Not enough slip will lead to poor machineability on the downstream converting equipment. Some slip may be absorbed by the adhesive system, which may weaken the inter-layer bond strength of the laminate.

The addition of 30-40% LLDPE to LDPE or EVA (4-5% VA) will improve seal strength and hot tack. For the highest hot tack and heat resistance, the LLDPE concentration should be increased to 70-80%. However, it is important to note that with linear rich blends, bubble stability will deteriorate thus making it more difficult to minimize film thickness variations. Dual lip cooling rings should be used with high LLDPE concentrations.

Figure 8.2 shows schematically the three types of air cooling ring designs commonly in use with polyethylenes. Type A is used with LDPE, but will reduce potential output from 2.3 to 1.5 kg/hr./cm with LLDPE. This design will lead to bubble instability with low melt strength LLDPE. Type B with dual orifice is recommended for LLDPE and will yield an output of 2.2 kg/hr./cm of die. The use of the air collar, which acts as a secondary cooling ring in C (stacked air rings), is recommended for heavy gauge films. This system will achieve an output of around 2.0 kg/hr./cm with LLDPE.

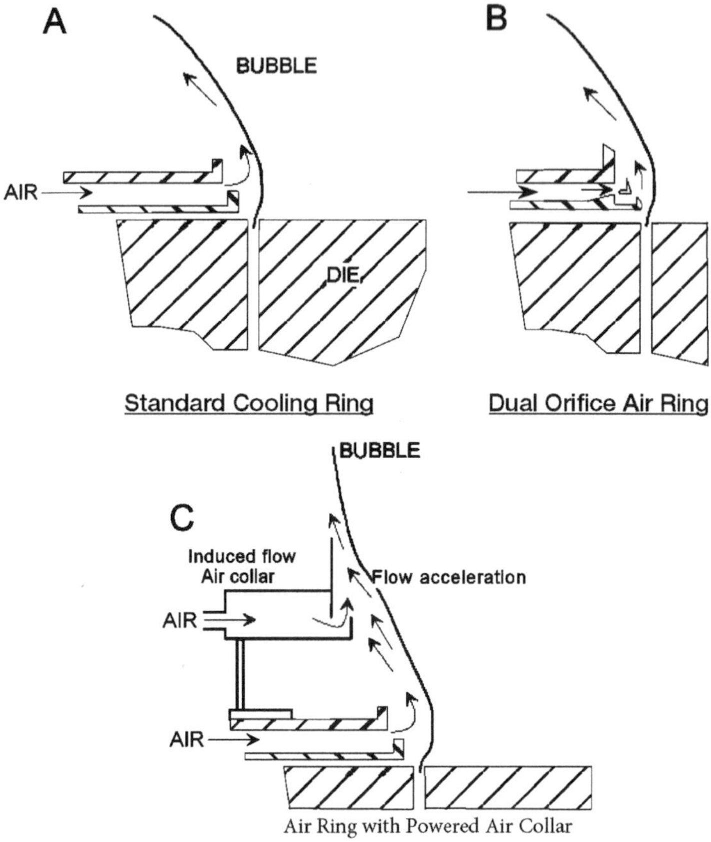

Figure 8.2

Cooling ring designs

Sharkskin will occur with LLDPE, if die gaps are not adjusted. **Figure 8.3** shows the effect on shear stress of five polyethylene types at three die gaps. Ziegler-Natta and metallocene LLDPEs are compared to LDPE. Resins with different comonomers (n-butene, n-hexene and n-octene) are included.

A shear stress of 140 kPa is generally accepted as the threshold for sharkskin formation. As shown in **Figure 8.3** the LDPE can be run with the preferred gap of 1.0 mm. The LLDPEs require die gaps well above 1.5 mm preferably in the range of 2.0-2.5 mm. The octene LLDPE (Z-N) develops the lowest shear stress of this group of linear PEs.

To optimize film properties balanced orientation in the machine (MD) and transverse (TD) direction is preferred. Orientation or stretch ratios are calculated as follows.

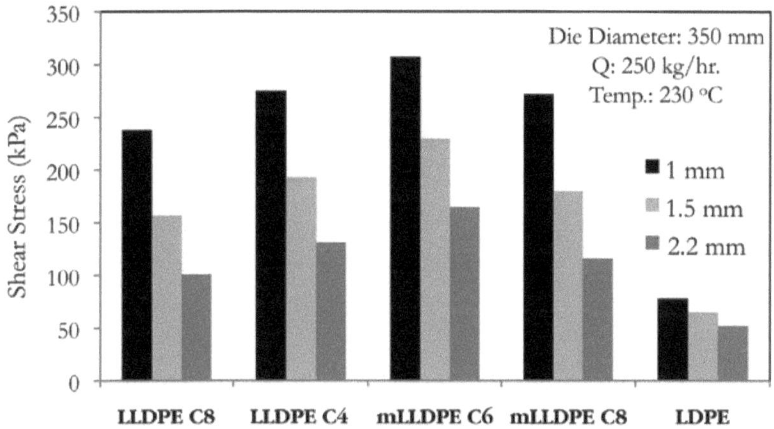

Figure 8.3

Effect of die gap on shear stress

The total stretch ratio (TS) from a tubular film process is the product of the MD and TD ratios shown in equation 8.1:

$$TS = MD \times TD \tag{8.1}$$

To obtain the TS the velocity at the die lips (V_1) relative to the line speed must be calculated. The velocity (V_1) is obtained from equation 8.2 (see section 1.6.):

$$Q = V \times t \times W \times \rho \tag{8.2}$$

The total stretch ratio (TS) is calculated as in equation 8.3 where V_2 is the line speed:

$$TS = V_2/V_1 \tag{8.3}$$

From equation 8.1 the MD stretch can now be calculated from the specified blow-up ratio (BUR).

Table 8.3 compares the draw ratios at two die gap settings. The line speed is fixed at 65 mpm and the BUR at 2 : 1. This example is based on a 300 mm diameter dic with a polyethylene throughput of 200 kg/hr. to produce a 30 μm film. The die gaps were set at 1 and 2 mm and the stretch ratios are calculated and shown in **Table 8.3**.

Table 8.3

Stretch ratios at 2 die gaps

Die gap (mm)	1.00	2.00
Velocity @ die lips V_1 (mpm)	4.65	2.33
Line speed V_2 (mpm)	65.00	65.00
BUR (TD stretch)	2.00	2.00
Total stretch ($TS = V_2/V_1$)	13.98	27.96
MD stretch	6.99	13.98

By doubling the die gap to 2.0 mm the MD stretch ratio is increased from 7 to 14. The tendency for the film to split and shrink in the MD is now increased very significantly. In addition the wider die gap will compromise the ability of the die to control film thickness distribution. The recommendation is to always minimize the die gap setting.

The development of high shear stress and pressure in the extrusion process will not only form sharkskin, but will also develop excessive temperature and motor load in the extruder.

A simulated experiment with a 90 mm diameter, 24 : 1 LD ratio barrier screw illustrates the processing differences between a mLLDPE (MI: 1.0 dg/min.) and an LDPE of similar melt index. The screw has a barrier gap of 1.27 mm and the clearance in the fluted mixer is 1.3 mm. The fluted mixer is a Maddock design with five channel pairs. The metering channel depth is 8.3 mm. The screw pitch in the barrier section is 1.3 D to accommodate the two flights and 1 D in the feed and metering zones. A schematic of the screw is shown in **Figure 8.4** with the temperature settings deliberately low. The screw was designed to minimize viscous dissipation. Output at 100 rpm is 233 kg/hr.

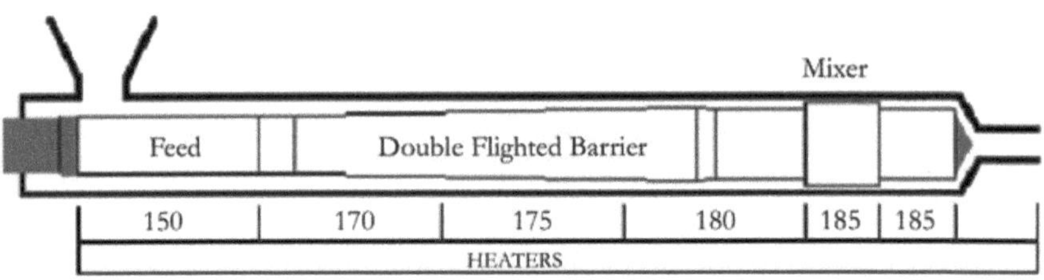

Figure 8.4

Schematic of barrier screw (diameter: 90 mm, LD ratio: 24 : 1)

The total power consumption for the mLLDPE = 97.25 kW and for the LDPE = 48.65 kW. The metallocene requires approximately double the power.

The melting process in the screw is plotted in **Figure 8.5**. The solid bed profiles along the screw are very similar for both polyethylenes. The rapid filling of the primary channel followed by rapid melting at 18 D illustrates the melting efficiency of this screw.

Figure 8.6 plots the temperature profile along the screw for both polymers. At a screw speed of 100 rpm the LDPE generates an exit temperature of 223 °C

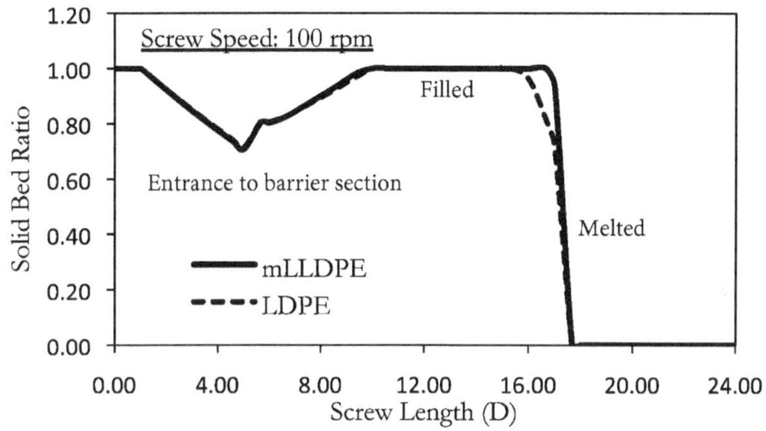

Figure 8.5

Melting in barrier screw

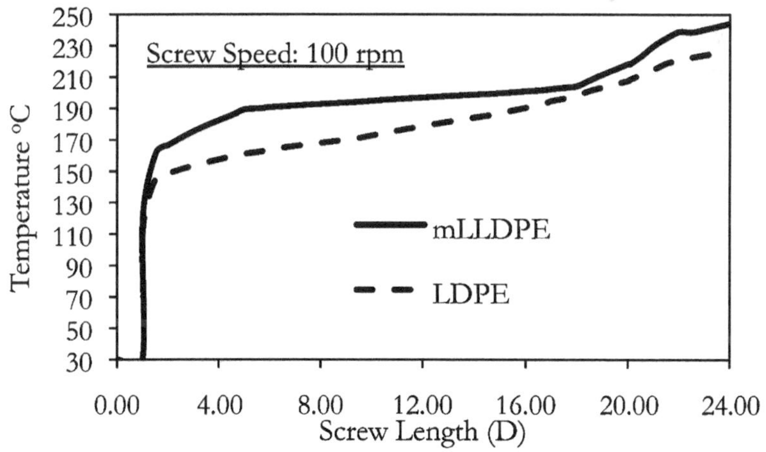

Figure 8.6

Temperature profile of mLLDPE vs. LDPE

compared with 239 °C for the metallocene. The mLLDPE exit temperature is 54 °C higher than the temperature settings (185 °C) in the metering zone of the extruder. The higher temperature will increase bubble instability. See chapter 13.10 for more information.

As a general comment setting a wide disparity between the barrel temperatures and the required melt temperature is not recommended. Melt temperature variations and output pulsation are more likely to occur. Ideally the process should be run close to isothermal conditions.

Processing aids (PPA) such as fluoroelastomers can reduce shear stress by generating slip along the die walls. These are dry-blended from masterbatches with the LLDPE and do not affect corona treatment. It is also claimed that these materials are effective in reducing die build-up and gel formation in polyethylenes. **Figure 8.7** shows computer generated data comparing the effects of slip at the die walls. The PPA does reduce shear stress at the walls, but is less effective than die gap enlargement.

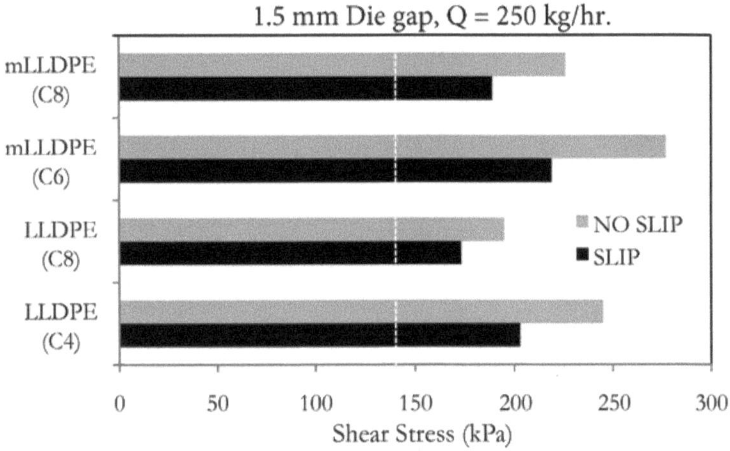

Figure 8.7

Effect of wall slip on shear stress of LLDPE

PPAs based on block copolymers of urethanes and polyols are also proposed as alternatives to the fluorinated polymers. These are lower in price and claimed to change boundary conditions at the die walls to generate slip and reduce shear stress. It is not clear how viable these products are compared to the fluoropolymers. The application of boron or titanium nitride surface treatments on the die lips can also reduce shear stress at the walls and is recommended by die manufacturers. Again the effect is marginal compared to die gap setting.

The easiest way to moderate shear stress is to blend with LLDPE or EVA (4-5% VA). **Figure 8.8** shows the effect of blending LLDPE with LDPE on shear stress at the die walls. To run with a 1.0-1.5 mm die gap the LDPE concentration must be at least 60%. This is defined as a linear lean blend. The throughput of the 350 mm diameter die is as above at 250 kg/hr. and 230 °C.

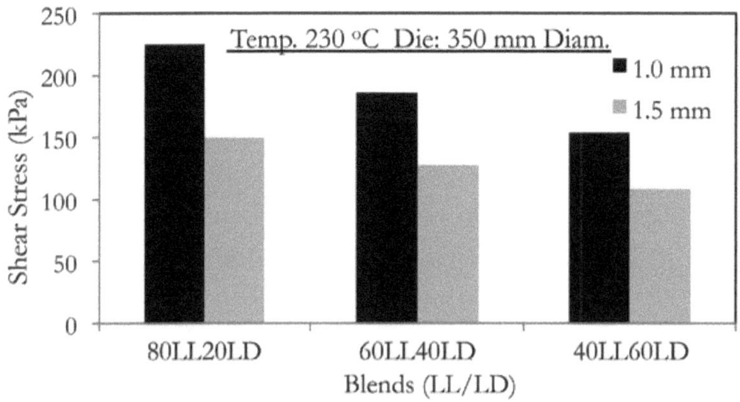

Figure 8.8

Shear stress with LL/LDPE blends at 2 die gaps

EVAs are used for low SIT (seal initiation temperature) and ionomers have outstanding hot tack, low SIT, excellent deep draw thermoformability and are able to seal through contaminated surfaces.

The **very low density** metallocene PEs (mPE) are excellent blending candidates for lamination films. Adding 15-20% of these metallocenes (density <0.910 g/cm^3) to LDPE will significantly improve hot tack, reduce SIT and lower haze. Haze and dart impact strength values comparing EVA (9% VA) and mPE (plastomer) are shown in **Table 8.4**. The outstanding dart impact is a result of narrow molecular weight and composition distributions. The extractables content is also very significantly reduced by the narrow composition distribution. This reduces film blocking and makes more efficient use of the slip agent.

Table 8.4

Haze and impact strength comparisons

	mPE	EVA (9% VA)
% Haze	3	3
Dart impact (g/mil)	>1050	200

Most film processors will have coextrusion facilities and will produce lamination films according to in-house recipes (see chapter 9). Coextrusion and polymer blending options are infinite with vigorous competition between producers (see section 9.13).

8.2 Corona treatment of lamination film

Corona treatment will increase the surface free energy of the film rendering it more receptive to wetting by liquids. It is important to remember that the plasma penetrates the surface to a depth that is unlikely to be greater than one micron. This means that the treatment can be rubbed off by friction. Loss of treatment in certain cases can be attributed to abrasion from stiff idler rolls on the converting machine. Softer films such as EVA copolymers are more prone to this rubbing off effect.

Elemental analysis of PE film surfaces by ESCA techniques confirm the presence of oxidized species following corona treatment. The data in **Table 8.5** was obtained on an LLDPE polymer free of slip/antiblock additives.

Table 8.5

Elemental analysis of treated and untreated PE film surfaces

	Carbon	O_2	Carbon	O_2
	Day 1		12 Months	
Not treated	99.2	0.8	99.4	0.4
Treated	86.9	12.2	91.0	8.6

The small amount of oxygen found on the untreated sample is probably due to the presence of impurities in the polyethylene. Treatment clearly increases oxygen concentration, which tends to drop slightly with ageing. However the film surface wettability did not change and the film was still printable.

In-line corona treatment during the extrusion process is preferred for two reasons. The first is that the more crystalline the polyethylene film becomes, the more difficult it is to treat. Since polyethylene films will increase in crystallinity on ageing, they will require more generator power after a day or two, for a required treat level.

Secondly, the films will usually contain a **slip** additive that will migrate to the surface to form a layer, which shields the polyethylene from the corona plasma. In **Figure 8.9** an LDPE film containing slip additive was corona treated in-line and the surface tension measured over a series of time intervals.

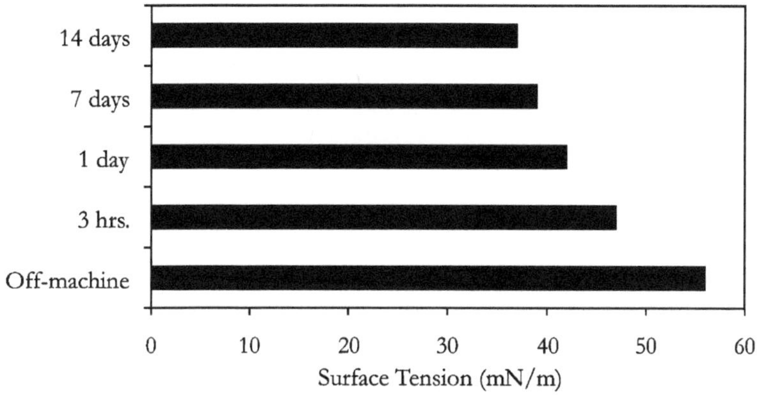

Figure 8.9

Effect of slip agent migration on treatment

Over a two week period surface tension drops from 56 to 37 mN/m. The biggest drop occurs within the first 24 hours. In **Table 8.6** the samples from **Figure 8.9** were retained and the surfaces washed with isopropanol. The surface tensions were re-measured. The initial surface tension is regained after washing.

Table 8.7 illustrates the difficulty of corona treating film that has been aged over a period of time. In these experiments the films were post treated one day and one week after extrusion. The lower treat levels achieved after the elapsed times, compared to the off-machine value, are caused mainly by the screening effect of the slip additive as it migrates to the surface. Increased crystallinity with time will also contribute to the lower treat levels. The drops are, however, not very significant after the first 24 hours.

Table 8.6

Effect of solvent wash on treatment (initial treat level= 56 mN/m)

Time	Unwashed (mN/m)	Washed (mN/m)
3 hours	47	56
1 day	41	56
7 days	39	56

The dependence of corona treatment on slip agent (erucamide) concentration is plotted in **Figure 8.10**. The slip level was varied from zero to a 1000 ppm in 50 μm thick film. The film line speed was fixed at 28 mpm and the corona generator was maintained at the same power setting throughout these experiments. As can be seen from **Figure 8.10**, treatment becomes more difficult as the slip agent concentration increases.

Table 8.7

Effect of time on post-treatment of film

Time	Treat level (mN/m)
Off-machine	56
1 day	50
7 days	49

As a conclusion it is important for the converter to select a corona treater of sufficient power for the throughput of the film line and ensure that the supplier provides the optimum design of electrode configuration. Additives, especially fatty acid amides and stearates will have a negative effect on treatability and these must be carefully controlled to avoid poor ink adhesion or inadequate bonding with lamination adhesives. Corona treatment should be applied before the additives have had time to migrate to the surface.

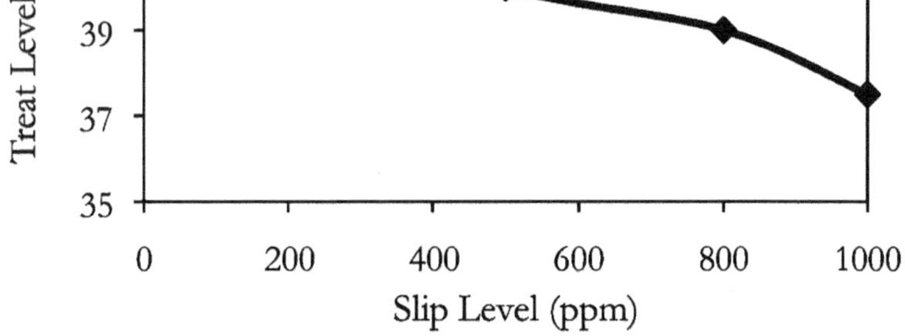

Figure 8.10

Effect of slip concentration on treatment level

Table 8.8 shows some recommended corona treatment levels with different types of printing inks on polyethylene films.

Table 8.8

Surface treatment levels for PE films

Printing process	Treatment (mN/m)
Flexographic/gravure	
Water	38-44
Solvent	36-40
UV	38-50
Litho	
Water	40-46
Solvent	37-42
UV	40-50
Offset/letterpress	
Water	40-46
Solvent	37-42
UV	42-54

8.3 Heat seal behaviour

One of the most crucial properties of the polyolefin layer in the laminate is its heat seal performance. Many choices are available to the laminator from simple LDPE and various types of LLDPE, EVA copolymers and ionomers. The metallocene PEs of different densities have further expanded the palette of choices.

In **Figure 8.11**, ethylene homopolymers and copolymers are compared for heat seal strength versus seal bar temperature (time 0.3 s.) The LDPE film requires a seal initiation temperature of at least 130 °C at these test conditions. The less crystalline VLDPE starts to seal at some 20 °C lower temperature, whilst the ionomer has the best response developing very high seal strengths at 110 °C. Above 150 °C the curve kicks upwards at these test conditions.

Polymerization of ethylene using metallocene single site catalysts produces polymers with both **narrow** molecular weight and composition distributions. This results in increased tensile strength and rapid crystallization rates. The outcome is films with higher seal strengths and faster hot tack. All the key polymer producers offer metallocene products that demonstrate improved heat seal performance.

Some heat seal data comparing two plastomer (mPE) type metallocene polymers to other sealant resins are shown graphically in **Figure 8.12**. The measurements were carried out on 2-layer 75 μm coextruded films of the following configuration:

HDPE 50 µm
SEALANT 25 µm

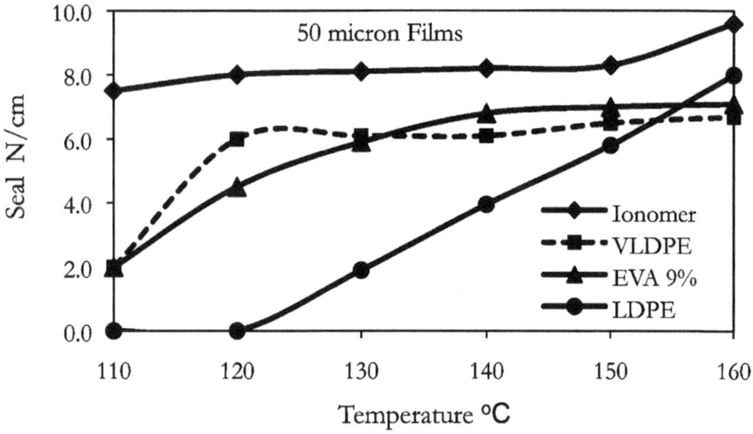

Figure 8.11

Seal strength of ethylene polymers

The basic characteristics of the products tested are summarized in **Table 8.9**. The mPEs as expected show greater structural homogeneity compared to the others.

The data in **Figure 8.12** shows the lower SIT of the mPE plastomers compared with LLDPE, ionomer and EVA. At 100-110 °C the curves begin to merge. However it is important to remember that there are other commercial ionomers which seal at lower temperatures than the product shown here.

Table 8.9

Properties of sealant resins

Sealant resin	MWD	CD	MI (dg/min)	Density (g/cm^3)	Comonomer
mPE #1	2 : 1	Narrow	1.2	0.900	n-butene
mPE #2	2 : 1	Narrow	1.2	0.900	Terpolymer
LLDPE	3-4 : 1	v. Broad	1.0	0.920	n-hexene
Ionomer	5-6 : 1	Broad	5.5	-	Zn^{++} salt
EVA	5-6 : 1	Broad	2.0	-	9% VA
CD: composition distribution					

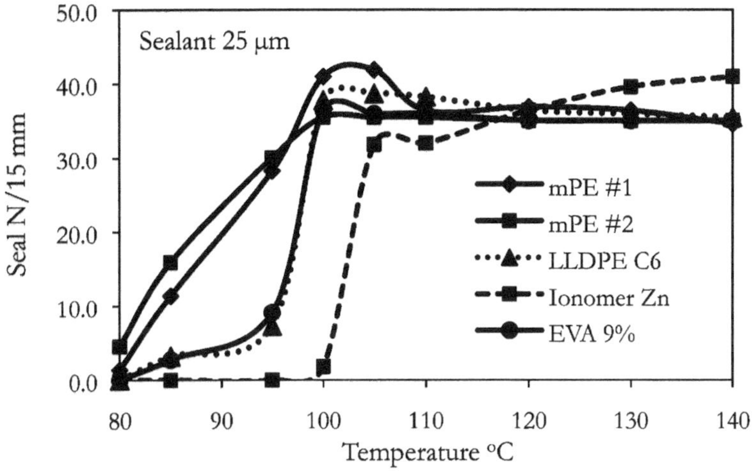

Figure 8.12

Comparative heat seal strengths of HDPE-sealant coextruded films

8.4 Hot tack

Hot tack performance has increased in importance as form, fill and seal packaging machines with faster output rates have evolved over the years. In this review hot tack will be described as follows:

Hot tack is the seal strength of the polymer above its melting point.

High hot tack polymers are those characterized by either high melt strength, such as ionomers, or polymers that exhibit a fast rate of crystallization. In both cases the hot polymer leaving the sealing jaws rapidly develops sufficient strength to resist the stresses induced by the applied force during filling of the pack. The hot tack data in **Figure 8.13** shows the superiority of ionomer and EAA (9%) over the other resins. The presence of hydrogen bonds in EAA, and ionic crosslinks plus hydrogen bonds in ionomers create a reinforcing network, which increases melt strength leading to much improved hot tack.

LLDPE crystallizes fastest and quickly builds up resistance to melt break, which results in somewhat improved hot tack as shown. The octene type LLDPE resins are believed to be superior in this respect to the lower α-olefin grades. Further improvements are claimed for the metallocene catalyzed copolymers.

Figure 8.14 compares the hot tack strengths of five coextruded films, which include two metallocenes. The sealant layer is 25 μm supported on 50 μm HDPE

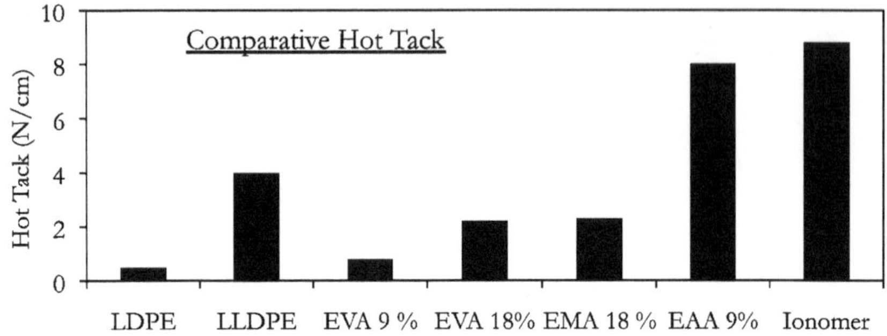

Figure 8.13

Hot tack of various PE resins

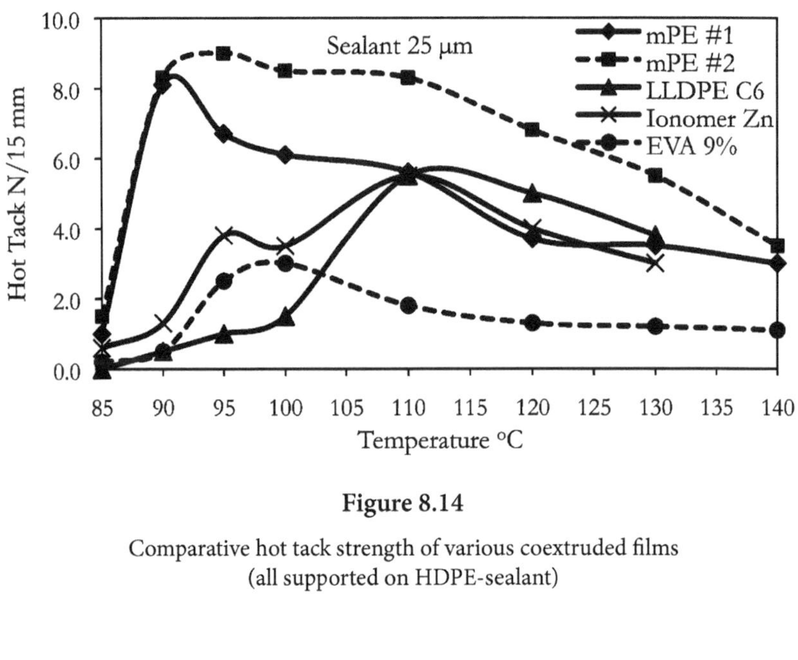

Figure 8.14

Comparative hot tack strength of various coextruded films
(all supported on HDPE-sealant)

as in **Figure 8.12**. The metallocenes particularly #2 develop high hot tack over a broad temperature range.

The effect of the adjacent layers on the heat transfer from the hot seal area will play a key role in the development of its hot tack strength. It is important to understand that the thermodynamic mechanism in hot tack is the reverse of sealing. In sealing, the critical step is the **heat transfer rate** through the web to the point of contact. In hot tack **removing heat away** from the contact point is critical.

Table 8.10 shows how the presence of other materials in a polyethylene based laminate will affect hot tack strength. In **Table 8.10**, the 60 micron LDPE has very low hot tack strength of around 1 N/15 mm of specimen width. Laminating the LDPE to polyester film (OPET) does not improve the situation since both materials have similar heat conductivity. With aluminium foil, hot tack improves dramatically as the metallic layer conducts heat away from the seal area. The last result in **Table 8.10** is interesting since the aluminium is now on the outside and the LDPE is in direct contact with the paper. The hot tack now drops to no more that 2 N/15 mm as the paper insulates the LDPE from the highly conductive metal.

This data illustrates that the performance of the seal layer will depend not only on its own rheological and morphological properties, but also on the other materials incorporated in the structure.

Table 8.10
Hot tack of different LDPE laminates

Laminate structure	Hot tack (N/15 mm)
LDPE (60 μm)	1.0
OPET (12 μm) /LDPE (30 μm)	1.0-1.5
Al foil (30 μm) /LDPE (30 μm)	7-10
Paper (50 g/m^2)/Al foil (12 μm) /LDPE (30 μm)	6-8
Al foil (8 μm) /paper (60 g/m^2) /LDPE (20 μm)	1.5-2

8.5 Additive migration to surfaces

If surface contamination such as a slip additive is present, intermolecular diffusion at the interface can be retarded resulting in weakened seals. **Figure 8.15** shows how a high concentration of erucamide slip additive in 100 μm thick LDPE film, can reduce heat seal strength and particularly its effect on the seal initiation temperature. The sealing time was 0.3 s.

Another phenomenon that will weaken seals is the presence of surface oxidation caused by intensive corona treatment. **Figure 8.16** shows how the seal initiation temperature is affected by a treatment level of 46 mN/m.

8.6 Surface contamination

Surface debris caused by contaminating the seal area during package filling can inhibit the integrity of the seal. In **Figure 8.17**, the seal strengths developed at a

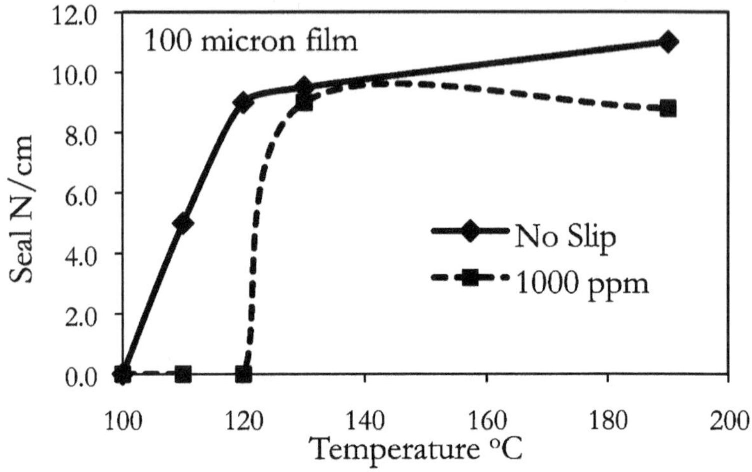

Figure 8.15

Effect of slip additive on seal strength

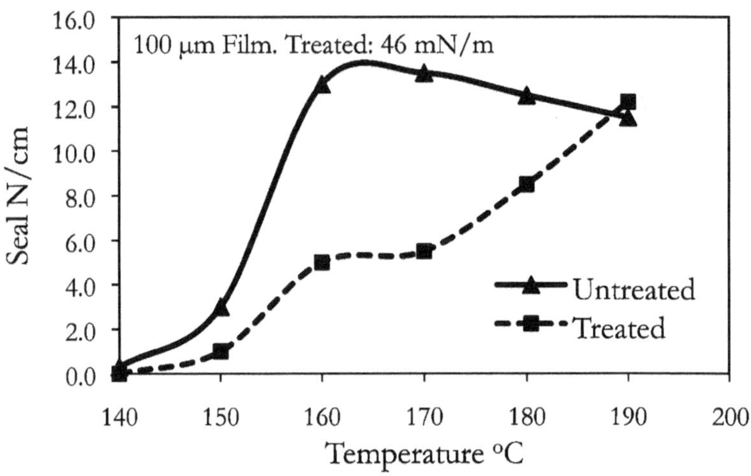

Figure 8.16

Effect of corona treatment on seal strength

bar temperature of 110 °C and a dwell time of 0.5 s are shown for some consumer products normally packed in aluminium foil-containing pouches. An LDPE (8 dg/min. MI) is compared to a zinc ionomer. Both polymers were coated onto an OPET/aluminium foil laminate. The coating weight was standardized to 50 g/m^2 and the line speed was 100 mpm. The ionomer was coated at 290 °C and the LDPE at 315 °C. The coatings were then smeared with the various consumer products and conditioned for 24 hours before sealing.

From **Figure 8.17**, the LDPE deteriorates in seal strength for all products except the curry powder. The uncontaminated sample (blank) develops a seal strength of 14 N/15 mm. Baking flour contamination seems to impair the sealability of both coatings the most.

Figure 8.18 shows the heat seal response of some extrusion coating resins in contact with bacon fat and milk powder. The EAA and the zinc ionomer come out best in these tests. The LDPE seal is very weak with bacon fat. The comonomer content in the EMA (20%) is approximately double that of the EVA and EAA. This high level of comonomer is unlikely to meet food packaging legislation for fatty products.

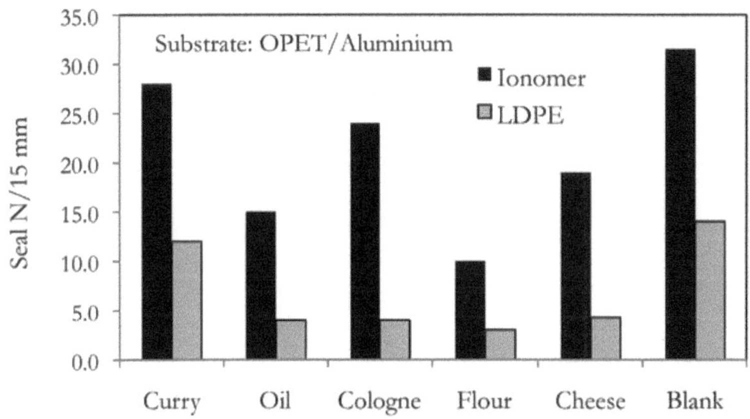

Figure 8.17

Effect of product on seal strength

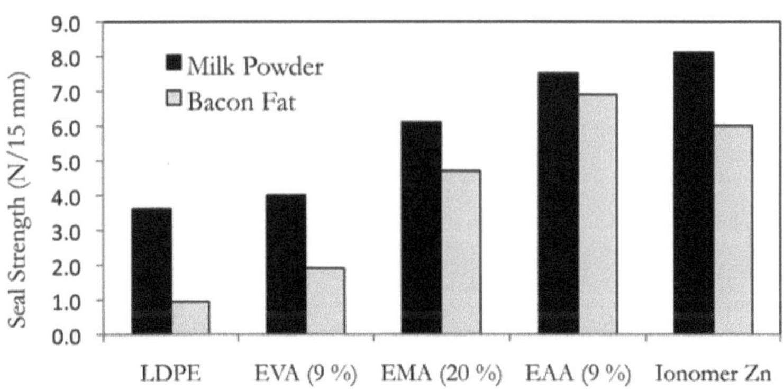

Figure 8.18

Seal through contaminants

8.7 Peelable seal film and coatings

The need for consumer friendly packaging has spurred the development of easy to open packaging. Peelable seals were developed to meet this need in flexible packaging. A number of approaches have been developed over the years to meet the conflicting requirement of easy openability with product protection.

Figure 8.19 illustrates the basic peel-seal mechanism. The films are sealed then pulled apart by the consumer to separate the sealant from the base stock. Another method is to propagate the peel by cohesive failure of the sealant (**Figure 8.20**). Delamination of the sealant layer from the base film stock is a third option.

Two basic techniques are used to apply the sealant layer.

1. Lacquer coatings, which can be separated relatively easily or broken down the seal.

2. Seal layers based on incompatible polymer blends, which can be coextruded, laminated or extrusion coated to the support.

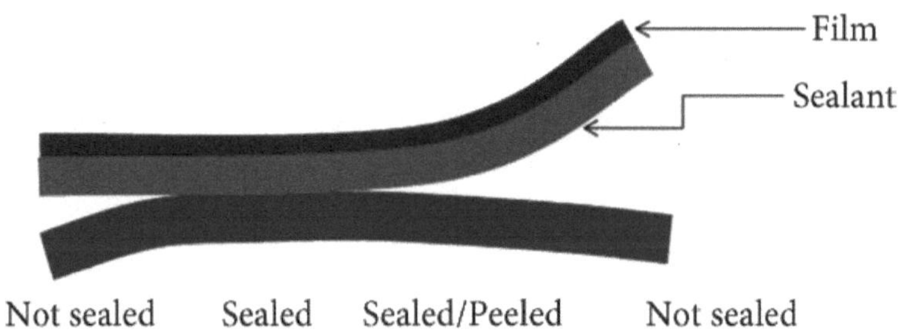

Figure 8.19

Seal-peel action

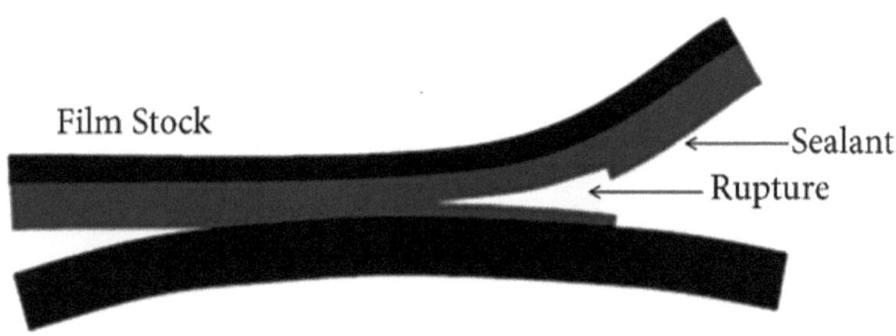

Figure 8.20

Sealant rupture

Lacquers are traditionally used as a peelable sealant for lidding stock based on OPET, BOPP or aluminium foil. Since the coatings are very thin, irregular distribution of the sealant will occur if the seal flange of the tray is not perfectly flat. This will lead to variable seal strengths, which may be overcome by higher sealing temperatures, but will tend to result in tear or break of the lidding stock support. Consequently other options have been developed over the years.

Peelable seals are now produced by modifying a resin's morphology with a material that inhibits the complete melting and fusion of its surface. This is usually accomplished by the use of incompatible blends. A binary blend is produced by mixing the base polyolefin resin with an immiscible polymer. **Figure 8.21** shows the non-miscible material forming a finely dispersed phase in a continuous phase of the host polymer, typically LL/LDPE or EVA. On heat sealing the continuous phase will seal as normal. The dispersed phase, however, forms a fracture prone interface, which will initiate separation or peeling when stressed.

Poly-n-butene (PB) is a useful polymer for the modification of EVA and LL/LDPE polymers. Another suggestion is the use of SEBS (hydrogenated SBS block copolymer.)

Obtaining the fine balance of seal integrity and easy peel is critical in not only the formulation, but also in the quality of mixing. For LDPE or EVA films a 15% addition of PB is usually recommended. LLDPE has a higher level of compatibility with PB compared to branched LDPE and higher concentrations may be required. Other solutions, such as blends of ionomers have also been proposed. Blending sodium and zinc ionomers in a 50/50 ratio is suggested and may prove simpler to control.

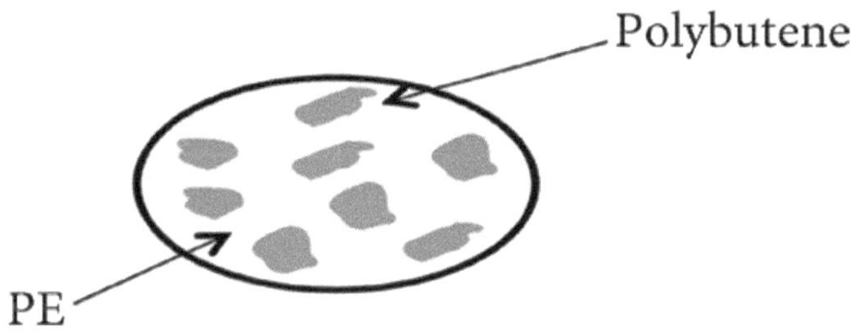

Figure 8.21

Incompatible blends

8.8 Scalping

It is important to be aware of the complex interactions that may occur between a food product and the laminate that it is in contact with. Any migration described as scalping of flavours and nutrients into the laminate will result in a loss of quality and should be taken into account when selecting materials and film thickness, **particularly the contact layer**, which is usually a polyolefin. Many aroma compounds, vitamins and nutrients present in natural foods can be transported into the amorphous phase of all polyolefins.

This scalping phenomenon is linked to the composition and morphology of the polymer, which determines its permeability (P). Permeability is a function of the solubility parameter (δ) and diffusion rate constant (D). The basic relationship is given in equation 8.4.

$$P = D \times \delta \qquad (8.4)$$

The absorption interaction or scalping constitutes the migration of specific organic compounds from the packed product into the polymer forming the contact layer.

The respective solubility parameters (δ) give a good indication of the mutual affinity of a polymer for a specified substance. The solubility parameter of a polymer, according to van Krevelen[1] is defined as the square root of the cohesive density in the amorphous state at room temperature. The constant is expressed as:

$$\delta = \sqrt{\frac{E_{coh}}{V}} \qquad (8.5)$$

E_{coh} is the cohesive energy of the substance and V is its volume per mol. The solubility parameter (δ) is expressed in J/cm^3 or cal/cm^3. **Table 8.11** shows the varying degrees of absorption (scalping) for a range of plastic films in contact with a variety of compounds.

Scalping by the two polyolefins is far higher than the polar polymers such as PET and polyamide. The polyolefins have solubility values of 15-16 J/cm^3 compared with 30-32 J/cm^3 for polyamide and EVOH. The values for the organic compounds are in the range of 14-20 J/cm^3 and closely matched to the polyolefins and are, therefore, more easily absorbed.

1 van Krevelen, D.W. and te Nijenhuis, K., Properties of Polymers, 4th Edition, **2009**, Elsevier Science, 978 0080548197, Ch. 7

Table 8.11

Scalping by polar and non-polar polymers

Compound		PE	PP	PVC	PET	PA (d/w)	PVDC	EVOH (d/w)
	δ (J/cm^3)	**15.6**	**16**	**19.4**	**21.4**	**30**	**24.4**	**32**
% Loss or scalping after conditioning								
Hexane	**14**	80	65	5	2	<1/3	<1	<1/1
Octane	**16**	75	60	10	2	1/3	<1	<1/2
Cyclohexane	**16**	70	60	2	1	1/2	1	1/3
D-limonene	**18**	70	65	15	4	1.5/5	<1	1/3
Citral	**18**	50	50	30	12	1/5	1	<1/5
Menthone	**18**	65	50	50	3	2/9	2	1/5
Ethyl acetate	**18**	35	30	45	8	4/5	1	2/10
Carvone	18	50	50	40	4	3/7	1	1/7
Anethole	**20**	35	30	75	25	1/3	<1	<1/7
Menthol	**20**	40	30	25	2	3/3	2	2/12
Benzaldehyde	**20**	45	40	60	15	4/11	2	2/15
Octanol	**20**	35	20	25	5	2/6	<1	<1/3
Me salicylate	**22**	40	35	20	7	5/7	1	3/12

d/w: dry/wet

A cocktail of organic compounds found in fruit juices and toiletry products are compared in **Figure 8.22**. The graph quantifies the much lower scalping levels of PET, PA and EVOH compared with the two polyolefins. Rigid PVC falls in between these two groups.

The key flavours isolated from toothpaste are shown in **Figure 8.23**. The bracketed numbers show the concentration of each component in ppm. **Figure 8.24** shows the scalping of a wide range of ethylene polymers in contact with the toothpaste over a 3-4 week period. As can be seen all these polymers scalp large amounts of these organic compounds. The data shows that there is no simple relationship with density or comonomer used.

This experiment confirms the conclusion drawn from the solubility parameter data that the **polyolefins** as a group will rapidly absorb compounds with similar δ values. The toothpaste flavours tested have δ values of 18-20 J/cm^3. The calculated δ values for the ethylene polymers in **Figure 8.24** lie between 16-18 J/cm^3. Clearly the migrating molecules are highly soluble in the polyolefins.

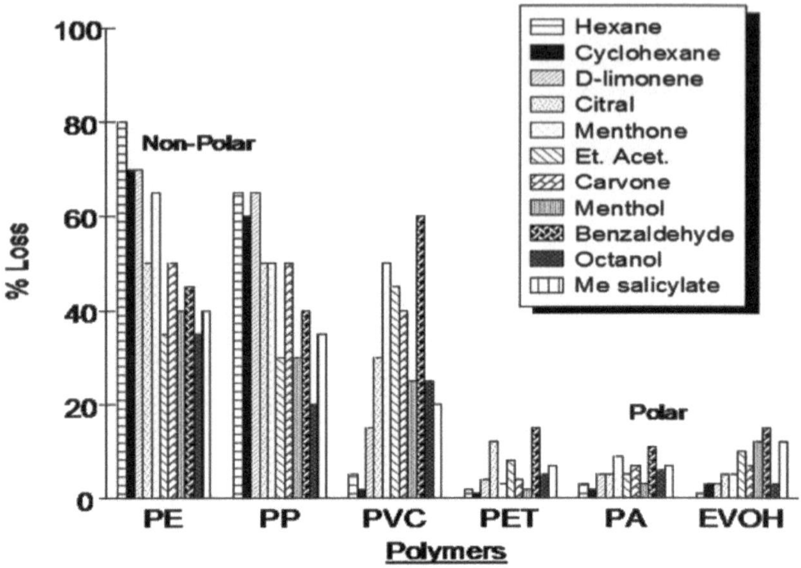

Figure 8.22

Comparative polymer scalping

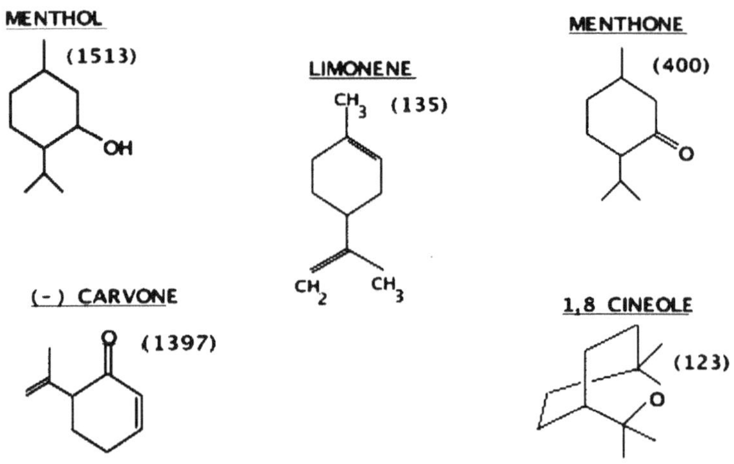

Figure 8.23

Flavour compounds in mint toothpaste, ( ) = typical ppm

Compounds with similar solubility parameters to those isolated from toothpaste are found in essential oils extracted from citrus fruits. These will pose a similar problem of flavour loss in packaging orange juice and similar beverages.

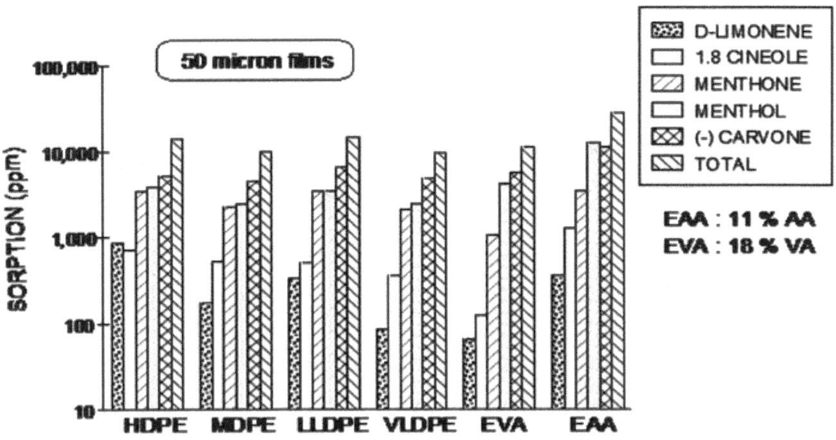

Figure 8.24

Scalping by ethylene polymers

Losses in flavour and nutrients are usually compensated for by artificially 'spiking' the packed product. Spiking requires specialized skills to reformulate the product by adding artificial ingredients and maintaining freshness. Another approach, limited in effectiveness, is to minimize the thickness of the polyethylene contact layer.

Chapter 9

Coextrusion processes

Coextrusion is the simultaneous extrusion of two or more polymers to form a multilayer structure. Laminar flow enables the separate melt streams to retain their integrity as they combine in the die. The process is used to produce thermoplastic composite films, sheet and coatings, which benefit from the synergy provided by all the component materials. By combining coextrusion with extrusion coating and film lamination virtually any product can be tailor-made to satisfy the needs of most commercial applications. The key is to determine the most cost effective solution.

9.1 Flat die coextrusion options

Coextrusion process options for the extrusion coater and cast film producer are shown schematically in **Figure 9.1(a)**, **(b)** and **(c)**. The most common configuration is the feedblock system (a) which is the lowest cost and the easiest method to accommodate two or more layers. The dual manifold dies (b) are costly to construct and more difficult to set-up, but offer distinct advantages. The melt streams are kept apart as shown and can combine either just inside the die or outside in the air gap. The two techniques can be combined by placing a feedblock module at the entry of one or both manifolds as shown in **Figure 9.1(c)**.

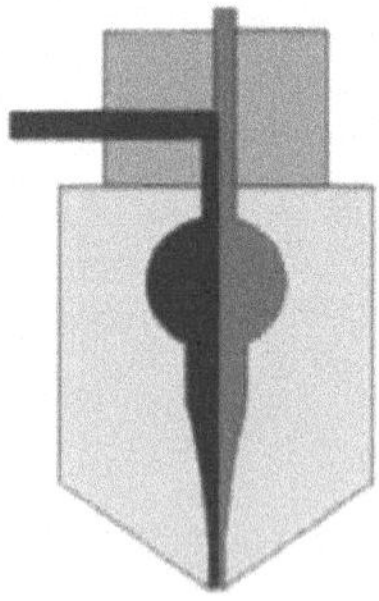

Figure 9.1(a)

Simple combining feedblock

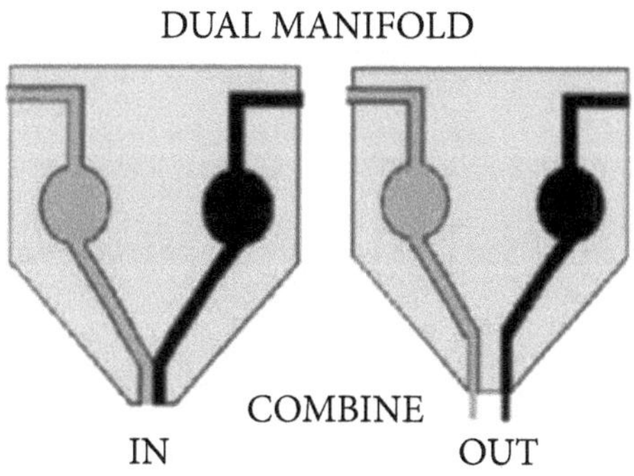

Figure 9.1(b)

Dual manifold options

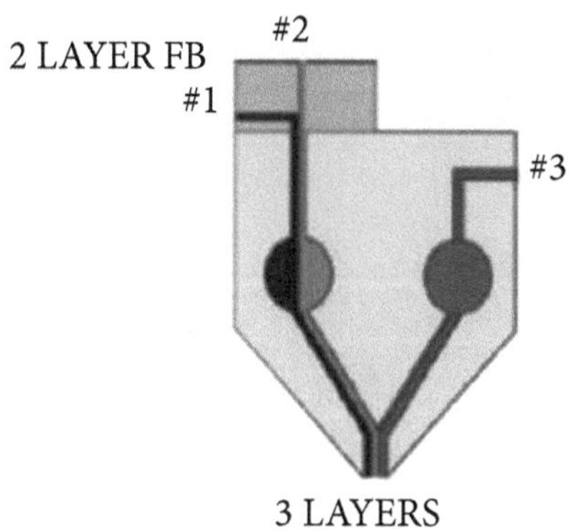

Figure 9.1(c)

Combination die

The advantages of coextruding with the combining feedblock include the following:

- Flexible: can handle many layers
- Allows encapsulation of corrosive polymers to protect die walls
- Skin layers easily carried on bulk layer
- Can be fitted to existing dies
- Cleaning relatively easy
- Single die adjustments
- Low cost

On the downside polymers with similar shear viscosities must be used and temperature differences between layers cannot be maintained.

Dual manifold dies are used in the following circumstances:

- When it is essential to keep the melt streams at different temperatures
- With materials of divergent melt viscosity

The operational difficulties with dual manifold dies relate to the need for two die adjustments and ensuring that the emerging melt streams have sufficient melt strength to avoid rupture or pinholing when drawn from the die exit. This will limit the minimum coating weight of expensive functional polymers. Another issue is high production downtime for polymer changes caused by readjustments of each die setting. The challenges are multiplied as the number of layers is increased.

The third option in **Figure 9.1(c)** combines the benefits of both systems and makes it possible to extrude three layers or more with a dual (2) manifold die. A second feedblock can be added to the free manifold.

Table 9.1 summarizes the key characteristics of the two coextrusion systems.

Table 9.1

Feedblock vs. dual manifold extrusion

	Dual manifold	Feedblock
Application to existing dies	Poor	Good
More than 2 layers	Poor	Good
Ease of thickness changes	Good	Good
Ability to change layer position	Poor	Excellent
Skin layer application	Fair	Excellent
Tolerance for corrosive polymers	Poor	Excellent
Maintenance of temp. differentials	Good	Poor
Tolerance to viscosity mismatches	Good	Poor
Ease of die adjustments	Poor	Good

9.2 Typical coextrusion problems

Practical experience over many years has demonstrated to the unwary processor that the melt flow behaviour of coextrusions can be unpredictable and much more complex than originally assumed. A number of problems can occur when two or more polymer layers are combined in a multilayer flow field. The most common defects are illustrated in **Figure 9.2**.

Poor layer uniformity is common in feedblock systems and is caused by the tendency of one layer to encapsulate another. The lower viscosity layer flows from the interface to the die walls and encapsulates the more viscous melt, which flows preferentially down the middle of the die. This phenomena is the most difficult to eliminate. As will be shown this problem is always present to some degree.

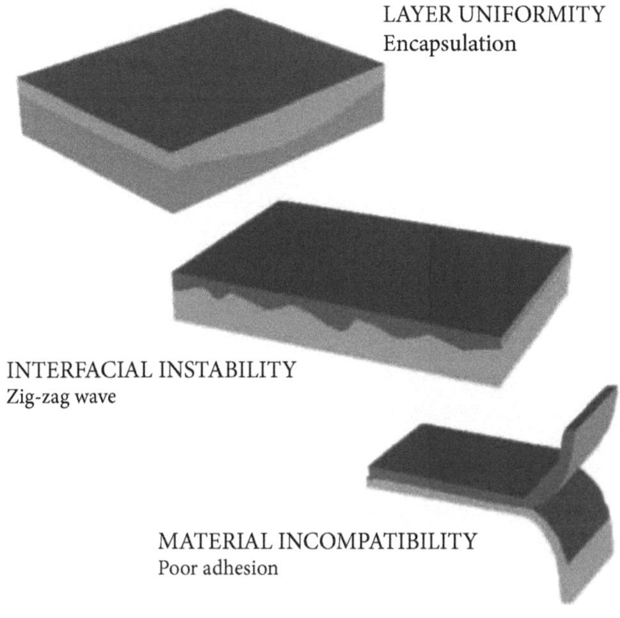

Figure 9.2

Common coextrusion defects

Interfacial instability occurs when excessive shear stress develops between layers. Zigzag wave instability similar to sharkskin can form. The easiest solution is to use low viscosity melts particularly close to the channel walls. Failing this option the phenomenon can be attenuated by geometric changes to the flow domain to reduce shear stress.

Material incompatibility results in poor adhesion and delamination at the interfaces. This problem can occur with all types of dies and processes and is usually resolved by the use of tie-resins, which act as mutually compatible bridges between the layers. In some cases blending of materials can enhance compatibility.

9.3 Layer rearrangement

There are three basic die shapes in extrusion processes:

1. Slit or slot dies
2. Cylindrical dies
3. Annular dies

In slit and cylindrical dies the layer interface will be in contact with the channel walls, which enables one layer to leak towards the walls. This will lead to layer rearrangement, i.e., encapsulation. In annular dies the interface is not in contact with the walls and rearrangement will not develop. Annular dies are used for pipe, tubing and blown film processes.

The tendency for a higher viscosity melt to encapsulate a lower viscosity fluid is demonstrated in **Figure 9.3**. The longer the flow path the greater the degree of encapsulation as illustrated by the five steps in these schematics. The lower viscosity layer (b) gradually envelops layer (a) as the melt progresses through the channel. A coaxial distribution can be observed with cylindrical dies when the LD ratio is >100 : 1. An energy minimisation principle has been proposed to show that the encapsulation of the high viscosity material by the low viscosity polymer results in a flow configuration that is the most efficient requiring the least total pressure[1].

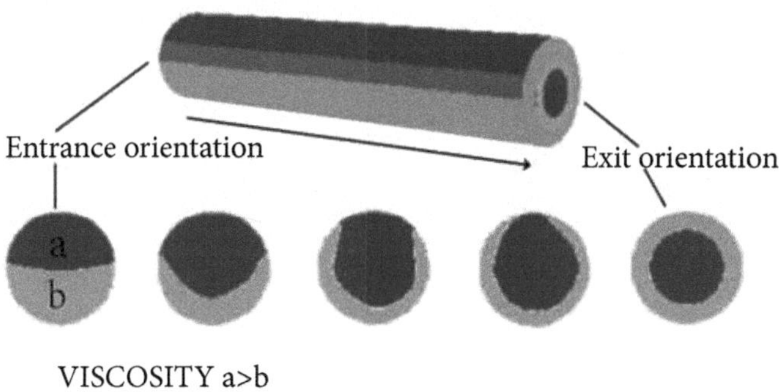

Figure 9.3

Encapsulation in coextrusion

This predicts the outcome, but does not explain when or where it will occur. The obvious solution is to keep the flow paths as short as possible. Using multimanifold dies where the melt streams are brought together just before the die exit or preferably beyond the exit is the best solution.

1 Karagiannis, A., Mavridis, H., Hrymak, A.N. and Vlachopoulos, J., Polymer Engineering & Science, **1988**, 28, p. 982–988

For many years the misconception persisted that interlayer thickness variation was caused entirely by viscosity differences. However, research over the years has demonstrated that the phenomena can occur with a single polymer in a coextrusion die. These studies suggested that layer rearrangement was also caused by secondary flows originating with viscoelasticity. These experiments demonstrated that highly viscoelastic polymers such as LDPE and polystyrene were more prone to rearrangement than polycarbonates. The problem is aggravated in long dies where elastic forces continue to affect structures throughout the flow domain. The phenomenon has been linked to the following properties and influences:

- Melt elasticity
- Normal stresses
- Memory
- Geometry

9.4 Coextrusion profiling

A PC simulation with a 1500 mm wide coat hanger die was used to illustrate layer rearrangement with two different material entries. The simulation model is based on shear flows and does not take into account viscoelastic response and is therefore limited in its predictions. Nevertheless the results are of interest and helpful in indicating solutions to the problem.

A 2-layer coextrusion with a pair of LDPEs of different melt index was run in this die. The die gap (lips) was set at 0.6 mm. The conditions are shown in **Table 9.2**.

Table 9.2

2-layer coextrusion conditions

Layer #1 LDPE	Layer #2 LDPE
MI 4.2 dg/min.	MI 8.0 dg/min.
200 kg/hr.	300 kg/hr.
40%	**60%**
Melt and die temperature: 310 °C	

The layer distributions shown graphically are expressed as percentages of the average volume flow rate (VFR) for each layer. The zero (0) position on the horizontal axis is the middle of the die. The half die width (750 mm) is plotted as shown in **Figure 9.4**.

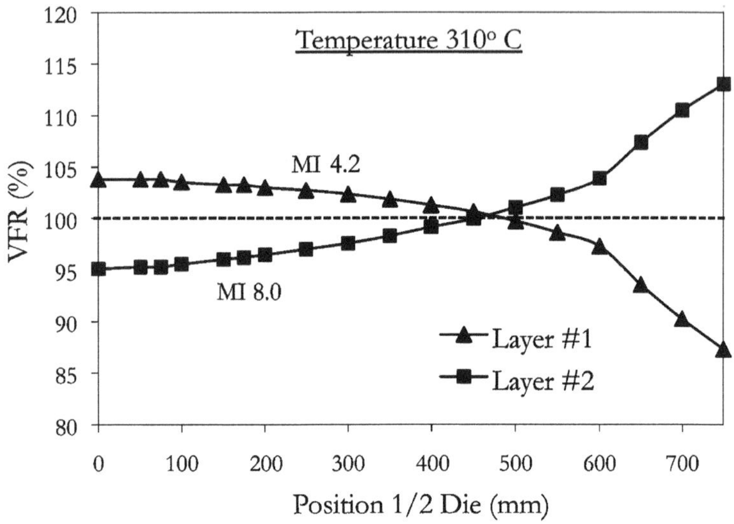

Figure 9.4

Layer distribution in 40/60% 2-layer coextrusion with different melt indexes (VFR = volume flow rate)

The graph shows that the 8 MI resin (#2) is pushed away from the middle towards the edge and begins to encapsulate the higher viscosity layer (#1), which flows preferentially through the middle of the die. The layer rearrangement is dominated by the viscosity difference.

A second experiment entered the 8.0 dg/min. melt index resin only, i.e. a perfect viscosity match, and the result is in **Figure 9.5**. This shows Layer #1 flowing preferentially towards the edges. Since this is the minor layer (40%) it is forced by the major layer (60%) away from the middle of the die towards the edges. Layer rearrangement is now dependent on the respective mass flow rates and **not** viscosity.

This example demonstrates two conditions in which layer rearrangement will occur, i.e., viscosity mismatch and mass flow rate ratio. Unfortunately, if one of the materials was run at 1 : 1 ratio, the software cannot differentiate between the two identical layers. In reality there may still be some layer rearrangement even with a perfectly matched viscosity and layer ratio that this computer model will not differentiate. However, the result does illustrate the problems that can occur and offers a useful method to test out options to minimize the effects of layer rearrangement.

Since there is no mathematical model or theory to predict how and when layer rearrangement will occur, die designers have over the years developed partial solutions to solve the problem. These solutions are based on a combination of intuition and experience. The technique most widely used is that of coextrusion profiling. This approach is to create an initial coextrusion shape that self adjusts to a uniform layer distribution as the melt stream flows through the die to the exit.

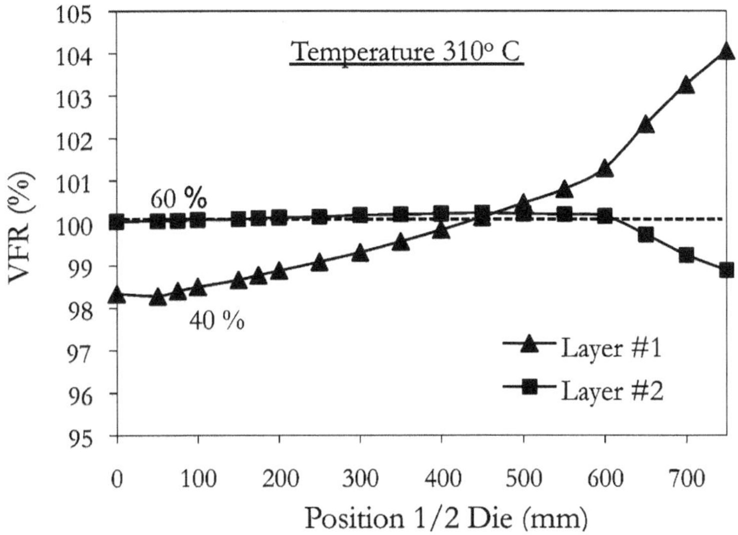

Figure 9.5

Layer distribution in 40/60% 2-layer coextrusion with matching viscosity

Figure 9.6 shows a typical feedblock profiling technique used to compensate for the encapsulation of the middle layer in this example. By profiling the middle layer into a concave geometry (bow tie) in the feedblock the encapsulation tendency by the outer layers is minimized.

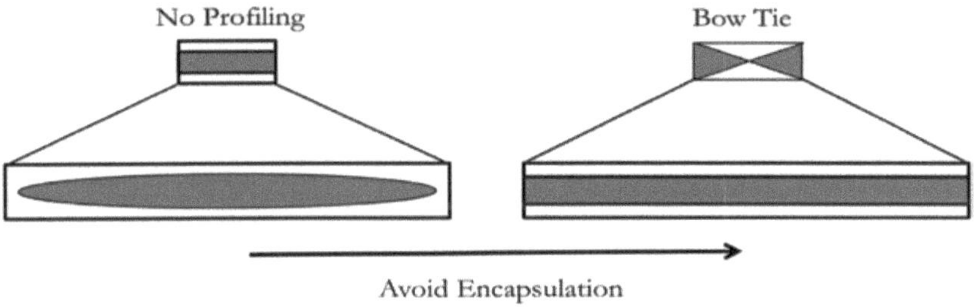

Figure 9.6

Feedblock profiling

Figure 9.7 is an example of diverting a low viscosity layer from the edges towards the middle. The diverter is placed in the feedblock and directs the lower viscosity layer towards the centre and the higher viscosity layer to the edges. This redistribution will ideally result in an even melt distribution at the exit. Modular feedblocks are designed with interchangeable inserts to divert the flow as required. Melt profiling is largely by trial and error and the problem becomes increasingly complex as the number of layers increases.

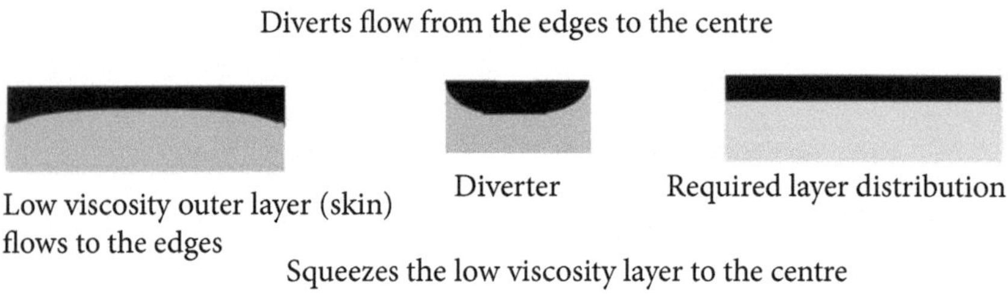

Figure 9.7

Typical profiling with diverter

9.5 Interfacial instabilities

In a simple coextrusion flow field as shown in **Figure 9.8** excessive shear stress at the interface can cause instabilities. The instability when it occurs will manifest itself by an increase in haze, as measured by narrow angle light scattering and the formation of zigzag waves. Reducing pressure in the die will usually eliminate these instabilities.

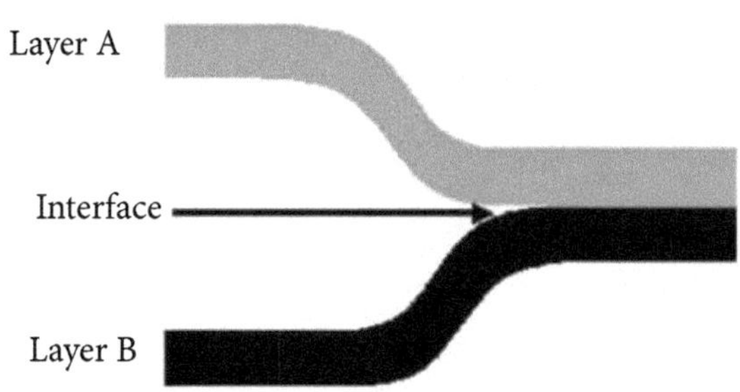

Figure 9.8

2-layer coextrusion flow field

Zigzag wave formations on a typical tubular blown film bubble are illustrated in **Figure 9.9**. Two types of wave can be formed as shown. The higher amplitude wave with a lower frequency can result in intermittent layer flow. Its origin is elusive, but could be initiated in the region where the polymers merge. Excessive melt acceleration at interfaces and the layer ratio can cause low frequency wave instability.

Two models are used to explain wave formation:

1. At a critical shear stress, interface adhesion drops causing slippage
2. Instabilities are caused by the development of **normal** forces at the interface

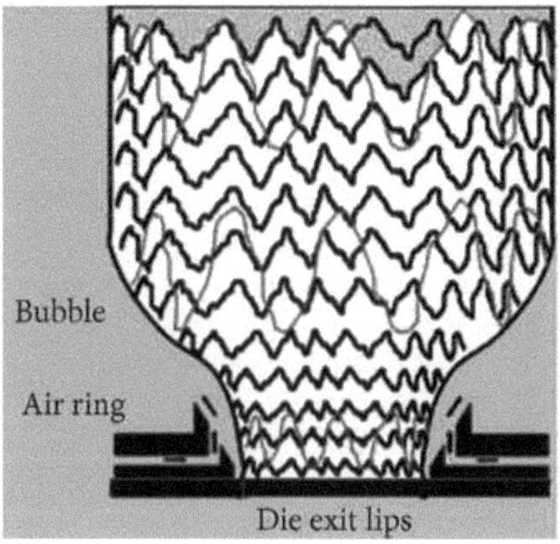

Figure 9.9

Waves and zig-zag waves

Irrespective of the model used observations have shown that the onset of interfacial instabilities can occur at a shear stress of approximately 30 kPa. If the shear stress reaches 50-60 kPa interfacial instability is highly probable. These critical values are subject to debate by theoreticians and should, therefore, be applied with some caution.

In a simple monolayer flow field shear stress will increase from the middle to the walls of the channel. An example is shown in **Figure 9.10**. The melt shear stress across a pipe of 4 mm diameter with a length of 25 mm is plotted. The LDPE is flowing at a rate of 500 kg/hr. at 300 °C. Shear stress is reduced from the wall to the centre of the flow domain. At the wall the shear stress reaches 140 kPa.

This observation must be considered in coextrusion where at least one interface will be present between layers. The closer the interface is to the die walls the higher will be the shear stress leading to possible instability. This can be problematic with outer skin layers where the interface is very close to the die walls. To minimize this effect a low viscosity melt should be placed as the outer layer closest to the walls. This will, however, encourage layer rearrangement as shown in **Figure 9.7** and a diverter will probably be necessary to equalize the flow across the die.

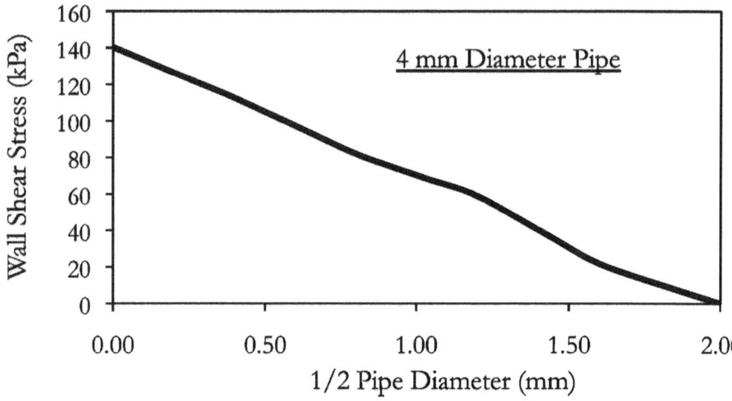

Figure 9.10

Shear stress from wall to middle

9.6 Computer simulations

A symmetrical tubular film coextrusion based on LDPE and HDPE was simulated. The final die section is configured as in **Figure 9.11**. The annular die diameter is 250 mm. The three layers are combined 10 mm from the exit and flow through a die gap of 1.5 mm. Melt and wall temperatures were fixed at 240 °C. The total mass flow rate was 185 kg/hr.

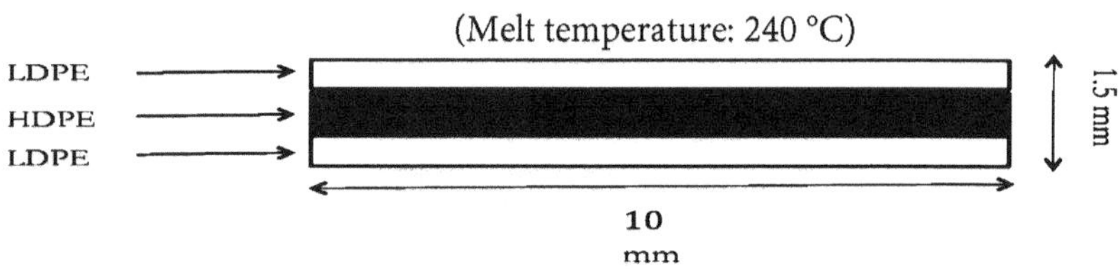

Figure 9.11

3-layer coextrusion with narrow gap

The basic properties and entered flow rates of the two polymers are shown in **Table 9.3**. The outer layers (LDPE) account for 10% of the thickness per side. The shear stress data at the walls and interface calculated by the software are summarized in **Table 9.4**. From these calculations the shear stresses are below the thresholds that are likely to lead to instabilities or melt fracture related defects.

Table 9.3
PE resin properties

	Layer #1	Layer #2	Layer #3
Polymer	LDPE	HDPE	LDPE
Input (kg/hr.)	18.3 (10%)	146.4 (80%)	18.3 (10%)
Melt index (dg/min.)	2.0	0.45	2.0
Density (g/cm^3)	0.923	0.945	0.923

Table 9.4
Interfacial shear stress (1.5 mm die gap)

Result (LDPE/HDPE/LDPE)	Value (in/out)
Shear stress at walls (kPa)	39.1/39.1
Shear stress at interfaces (kPa)	26.6/26.6

The LDPE is now replaced by a plastomer type metallocene polyethylene in order to provide superior heat seal properties. This resin has a melt index of 1.1 dg/min and a density of 0.902 g/cm^3 and is characterized by very narrow molecular weight distribution. Mass flow rates and temperatures were unchanged. The results in **Table 9.5** show that much higher levels of shear stress have now developed inside the die.

Table 9.5
Shear stress in die (1.5 mm die gap)

Result (mLLDPE/HDPE/mLLDPE)	Value (in/out)
Shear stress at walls (kPa)	151/151
Shear stress at interfaces (kPa)	100/100

Shear stress of 151/151 kPa at the walls will possibly result in sharkskin and 100 kPa at the interfaces will lead to severe interfacial instability. Redesign of the die is now required in order to run this coextrudate. The simplest option was to widen the die gap to 2.5 mm. The resultant shear stress data is summarized in **Table 9.6.**

The wall shear stress is reduced to 82.8 kPa and is well below the threshold for the onset of sharkskin. However the interface shear stress is still too high at 54.9/54.9 kPa.

Table 9.6
Shear stress in die (2.5 mm die gap)

Result (mLLDPE/HDPE/mLLDPE)	Value (in/out)
Shear stress at walls (kPa)	82.8/82.8
Shear stress at interfaces (kPa)	54.9/54.9

The next step was to increase the ratio of mLLDPE to HDPE in the system. This will push the interface away from the walls to a region of lower shear stress. Total mass flow rate was kept at 185 kg/hr. and the outer layers were increased to 15% per side as in **Table 9.7**.

Table 9.7

PE resin properties

	Layer #1	Layer #2	Layer #3	Total
Polymer	mLLDPE	HDPE	mLLDPE	kg/hr.
Flow Rate (kg/hr.)	27.75 (15%)	129.5 (70%)	27.75 (15%)	185

Table 9.8 shows that moving the interface away from the walls by increasing the thickness of the mLLDPE outer layers reduces the shear stresses at the interfaces. The 44 kPa interface shear stress is now borderline.

Table 9.8

Interfacial shear stress (2.5 mm die gap)

Result (mLLDPE/HDPE/mLLDPE)	Value (in/out)
Shear stress at walls (kPa)	79.0/79.0
Shear stress at interfaces (kPa)	44.0/44.0

The simulation clearly shows the difficulties that may be encountered with the use of very narrow molecular weight distribution (MWD) polymers. Widening the die gap further is not recommended since at 2.5 mm the gap is already very wide and will lead to excessive drawdown in the machine direction (MD) and also result in poorer control of thickness distribution from the die. Increasing melt temperature will make it more difficult to cool the bubble and will reduce melt strength. Reducing the throughput of the die will reduce pressure, but will obviously result in productivity loss. The most practical solution is to blend the mLLDPE with LDPE or EVA, which will significantly reduce the overall pressure in the die. Another option, which adds cost, is to blend with fluoroelastomer processing aids. These will generate slip at the die walls thus reducing pressure (see section 8.1).

Coextrusion can be combined with extrusion coating of a substrate to provide more complex structures.

In these systems where high temperatures and high melt index resins are used pressures in dies are relatively low and excessive shear stresses are avoided. However, thin skin layers are also very often used where the interfaces will be close to the die walls where stress is highest. The effects of the proximity of the interface to the die walls on thickness distribution and shear stress are evaluated below. Simple coextrusions typically used by extrusion coaters were simulated using the die model in **Figure 9.12**.

The die width was 1500 mm with a final exit gap of 0.6 mm. The land length was 10 mm. The coat hanger die is assumed to be fed from a feedblock unit. Wall temperatures are 305 °C. The flow distribution and interfacial shear stress were calculated relative to the position of the interface in a simple 2-layer EAA/LDPE structure widely used in extrusion coating. The key parameters are entered in **Table 9.9**. The EAA layer thickness was varied from 50% to 5% of the total mass flow rate.

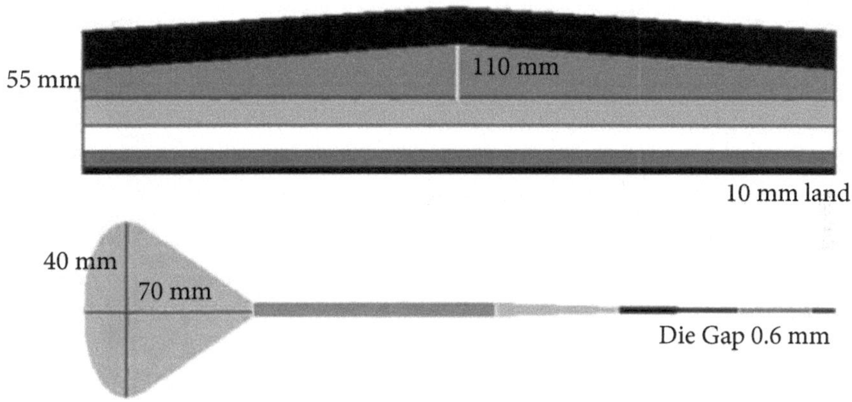

Figure 9.12

Coat hanger die design used in simulations

Table 9.9

EAA/LDPE coextrusions

Structure	Flow Rate (kg/hr.)	% Layer	Temp. (°C)
EAA/LDPE #1	250/250	50/50	295/310
EAA/LDPE #2	50/450	10/90	295/310
EAA/LDPE #3	25/475	5/95	295/310
LDPE MI 8.0 dg/min. EAA MI 8.0 dg/min. (6% AA)			

In **Table 9.10** the shear stresses developed at the EAA/LDPE interfaces are calculated. As the interface moves from the middle towards the walls of the flow field the shear stress increases from 5.6 to 30 kPa. In this example the highest shear stress developed is unlikely to result in any instability. However, in processes where lower temperatures and more viscous polymers are used instabilities may occur.

Table 9.10

Moving EAA/LDPE interface closer to walls

% EAA/% LDPE	Shear stress at interface
50/50	5.6 kPa
10/90	28 kPa
5/95	30 kPa

The thickness distribution expressed as percentage of the target values are plotted in **Figure 9.13** for the three layer ratios. The curves show that as the EAA layer becomes thinner the melt stream is forced away from the centre towards the edges. The best result is with the interface (50/50) in the middle of the channel. The concern with skin layers is whether this transfer of material to the edges is acceptable for commercial use or needs to be corrected with a diverter in the feedblock. The thicker edges will usually be trimmed by the slitter knives on the coating machine.

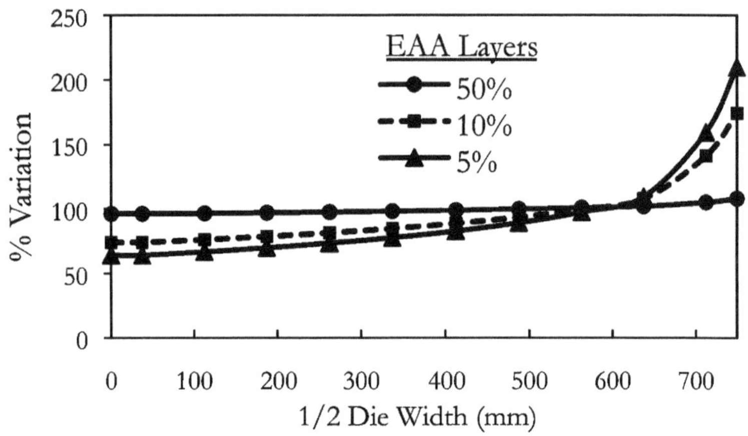

Figure 9.13

Effect of EAA layer thickness

9.7 Some methods to minimize interfacial shear stress

Three easily controlled options are considered for their relative effectiveness in moderating shear stress. These assume that the use of lower shear viscosity polymers and layer positioning is proscribed.

1. Adjust the die gap
2. Increase temperature
3. Vary total output

A simple symmetrical 3-layer structure is used as an example. The geometry and material inputs are shown below. Metallocene PE (mPE) with LDPE was selected as the worst case scenario.

- Die width: 1500 mm
- Land length: 10 mm
- Die gap varied: 0.6 to 1.5 mm
- Mass flow rate: 500 kg/hr.
- Structure: 10% mPE/80% LDPE/10% mPE

Figure 9.14 shows the interfacial shear stresses with this 3-layer symmetrical coextrusion with different die gaps at 310 °C. At a die gap >0.8 mm shear stresses are below the 30 kPa threshold. Clearly 0.6 mm is too narrow a die gap. A gap of 1.5 mm is probably excessive. The optimum die gap is in the range of 1.0-1.2 mm.

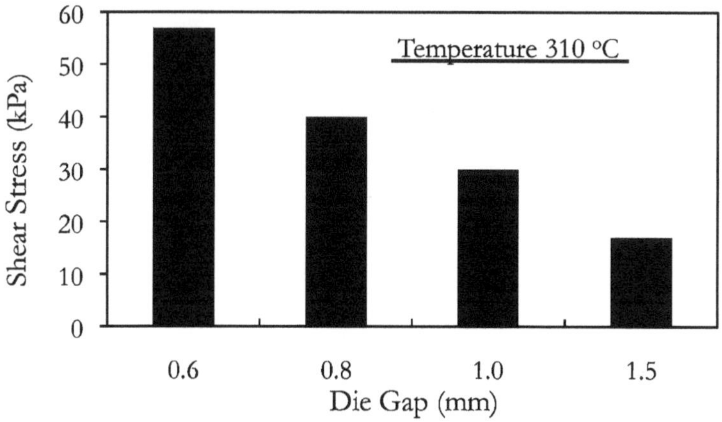

Figure 9.14

Effect of die gap opening on shear stress

Figure 9.15 shows the effect of temperature on shear stress. Over a temperature range of 300° to 315 °C the reduction in shear stresses is minimal. All the values are well above 30 kPa.

In **Figure 9.16** the effect of reducing mass flow rate from 500 to 400 kg/hr. on shear stress is calculated. Again the change is minimal. The latter approach is undesirable as it will reduce productivity for little gain. These calculations show that the geometry of the flow domain has the most significant effect on the development of shear stress at interfaces. In other words opening the die gap is usually the best and easiest solution to overcoming excessive shear stress situations. The assumption here is that the problem occurs only at the die lips. However, there is always the possibility that this phenomenon is initiated further upstream in the die and die lip adjustments may have little impact. In these cases the remedy becomes much more complex.

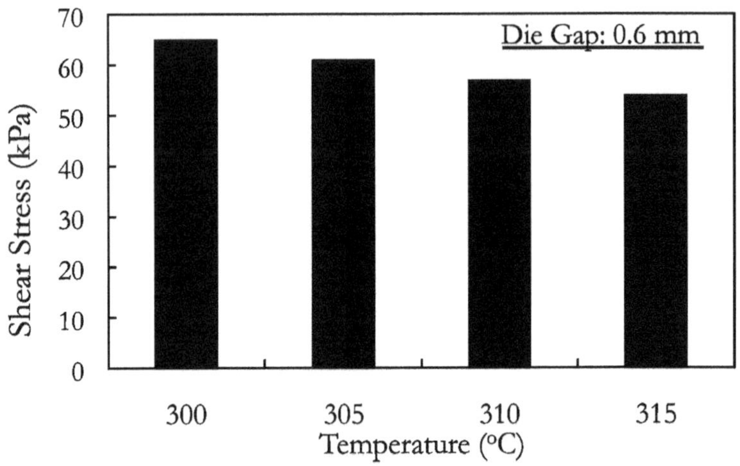

Figure 9.15

Effect of melt temperature on shear stress

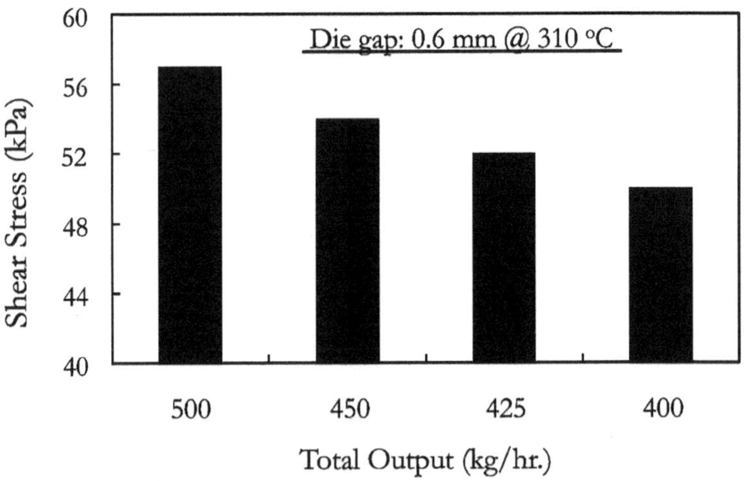

Figure 9.16

Effect of mass flow rate on shear stress

9.8 Flow dividers

Interfacial instabilities can also be caused by large differences in velocity of melt streams as they combine inside the feedblock. A method to eliminate this problem is the use of flow dividers. One method of adjusting channel geometry is with moveable vanes or flow dividers inside the feedblock. The externally adjustable vane is rotated about its axis and consequently changes the exit area, and hence the velocity in the flow channel. The intent is to adjust the velocities of each layer so that at the point of intersection the velocities are matched, thus reducing excessive elongation or stretching at the interfaces.

The usefulness of flow dividers in adjusting velocities at interfaces in a coextrusion can be assessed by computer simulations. An LDPE and EAA copolymer with the characteristics shown in **Table 9.11** were inserted in a 2-layer flow domain. The EAA mass flow rate was 10% of a total of 500 kg/hr. Wall and melt temperatures were 295 °C.

Table 9.11

Polymers used in coextrusion simulation

Entry conditions	LDPE	EAA
Melt index (dg/min.)	8.0	8.0
Density (g/cm^3)	0.917	6% AA
Mass flow rate (kg/hr.)	450	50

An expanded view of the channel as the two layers combine is shown in **Figure 9.17**. A streamline originating near the EAA wall where the layers combine is also shown. An EAA particle moving along this streamline will accelerate from a velocity of

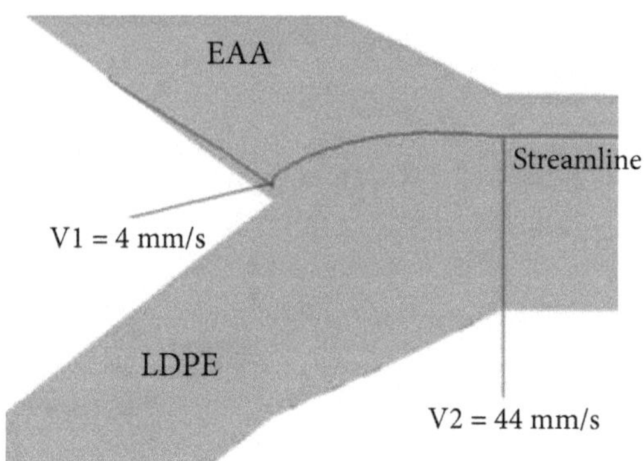

Figure 9.17

Velocity streamline at interface

4 mm/s to 44 mm/s as it combines with the LDPE flowing at a much higher velocity. This high velocity ratio of ~10 : 1 may cause wave instability at the interface.

To reduce this velocity difference a flow divider is introduced in the EAA channel. In the simulation this is configured as an internal dam and shown in **Figure 9.18**.

The velocity of the EAA particle has now increased to 32 mm/s past the divider and remains virtually unchanged at 30 mm/s as it combines with the LDPE and enters the die. The velocity ratio has now been reduced to approximately 1 : 1.

This PC simulation shows how to control velocity by adjusting the geometry of the internal channels to overcome excessive accelerations at layer interfaces. The elongational viscosity of the polymer can play an important role in the outcome, particularly with strain hardening polymers such as LDPE where high stretching stresses may be generated.

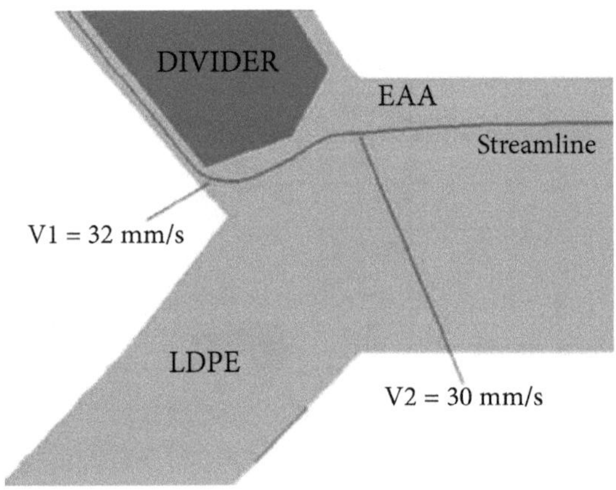

Figure 9.18

New velocity streamline at interface

As a summary in coextrusion coating with LDPE and ethylene copolymers pressure in the die will be low and instabilities are rare. However, with higher shear viscosity polymers such as polyester and EVOH at lower temperatures, die pressures can be much higher and flow instability may occur.

9.9 Heat transfer inside a coextrusion die

Combining different materials in a multilayer structure will often give rise to problems associated with different melt temperatures. For example many polymers will not tolerate the temperatures typically used for LDPE in extrusion coating. EVOH can degrade at temperatures >240 °C and extrusion temperatures should be set according to the suppliers recommendations. The EVOH melt stream will usually enter the die via a feedblock where it will combine with another material probably set at a higher temperature. The imperatives of good adhesion require the melt to be brought into contact with the substrate, e.g. paper, at high temperature. A simulation was carried out to predict the heat exchange in the die with the polymer melts entered at different temperatures. A symmetrical 3-layer PE/EVOH/PE coextrusion was evaluated.

The 1500 mm wide coat hanger die used previously (**Figure 9.12**) was entered in the program. The inputs to the process are summarized in **Table 9.12**. The EVOH is entered at 260 °C and the LDPE at 305 °C. Die body temperature was also set at 305 °C. With the EVOH representing only 10% of the mass flow rate the two interfaces will form in the middle of the flow field, i.e., away from the walls, and interfacial shear stresses are expected to be low. The simulation yields interfacial shear stresses of approximately 5 kPa.

Table 9.12

LDPE/EVOH/LDPE coextrusion

Polymer	Mass flow rate (kg/hr.)	Distribution (%)	Temp. (°C)
LDPE (3.5 dg/min. MI)	225	45	305
EVOH (high C_2)	50	10	260
LDPE (3.5 dg/min. MI)	225	45	305

The heat transfer between the three layers was calculated as the combined melts flow through the die. **Figure 9.19** plots the melt temperature from the entry port to the die exit.

In **Figure 9.19** the two LDPE layers enter the die at 305 °C combined with the EVOH at 260 °C. As the melts flow through the die the EVOH temperature is raised and exits the die at just above 306 °C. Some viscous dissipation occurs. The effect of the two major LDPE layers is to overwhelm the final average temperature by the time the melt reaches the die's exit. This simulation demonstrates the difficulty of trying to maintain a temperature differential between layers particularly where there are wide differences in thickness ratios. In this example the heat sensitive EVOH is buried between the inert polyethylene and not in contact with the die walls. The possibility of degraded EVOH building-up on the die walls is minimized. However, unless completely encapsulated, the EVOH layer interface at the edges will be in contact with some parts of the die surface. This may lead to some degraded material build-up at the walls. Once this happens it becomes very difficult to avoid the formation of gels and other melt defects.

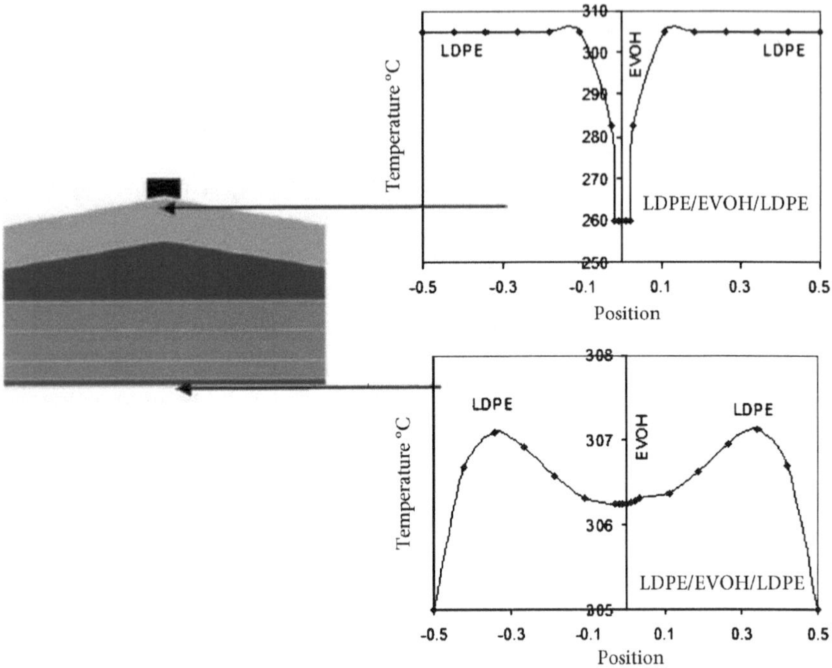

Figure 9.19

Heat exchange in a 3-layer coextrusion

9.10 Coextrusion with dual manifold dies

Some die designers claim that dual manifold designs can maintain temperature differences of some 20 °C between melt streams. These were first used to extrusion coat two layers of LDPE at high and low temperature. The layer facing the substrate is the so-called 'hot' layer to maximize adhesion. The outer 'cold' layer reduces the undesirable side effects of high oxidation, which leads to off-taste and odour in sensitive food packaging. A PC simulation was carried out to estimate the heat transfer between two layers of LDPE fed at different temperatures into a dual manifold die. The calculation is planar across a 1500 mm wide die and the inputs are in **Table 9.13**.

Table 9.13

Polymer input in coextrusion die

LDPE layers	Flow rate (kg/hr.)	Melt temp. (°C)
Layer #1 (cold)	425	295
Layer #2 (hot)	75	315
Wall Temp. (°C)	Layer #1: 295 and #2: 315 °C	

The high temperature oxidized polymer layer is 15% of the total flow rate of 500 kg/hr. In **Figure 9.20** the channel where the two layers combine is minimized to 1.0 mm in length. The die gap is 0.7 mm. The lip temperature is set at 305 °C. The temperature profile across the 0.7 mm die gap is plotted in **Figure 9.21**. The bulk layer temperature is mainly 295 °C and the peak temperature of the minor layer has only dropped from 315 to 311 °C. This should be sufficient to generate acceptable adhesion to the substrate.

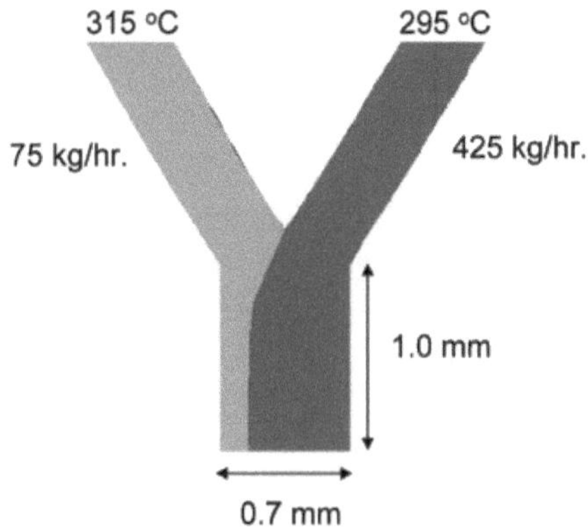

Figure 9.20

Die lip section of dual manifold die

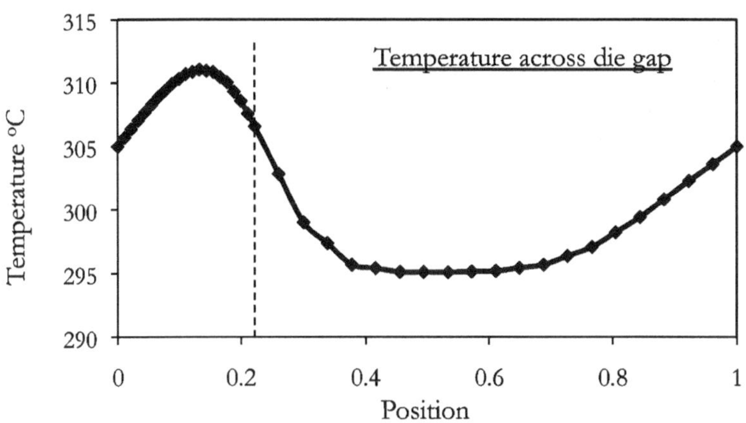

Figure 9.21

Temperature across very short land

These simulations show that it is possible to maintain an approximate temperature gradient of 15 °C providing the contact times are minimized. The channel where the layers are in contact should be of the order of 1-2 mm in length or combined outside the die. This is particularly important when there is a large difference in mass flow rates between layers. The assumption in these calculations is that there is sufficient insulation between the two manifolds as the melt streams flow through the body of the die.

9.11 Coextrusion coating with ethylene copolymers

Combining a low cost polymer such as LDPE with an extrudable adhesive has led to the rapid growth of coextrusion coating as a way to increase productivity and improve adhesion. The combining feedblock is the preferred process. Since aluminium foil is the best barrier material for the coater this has led to extensive evaluations over the years comparing the product resistance of different coextrusion adhesives coated on foil bearing substrates.

The functions of the coextrusion adhesive are:

1. To adhere to the substrate in this case aluminium foil
2. To adhere to the polyethylene support layer
3. To reduce melt temperature
4. To resist attack by the packed product during its shelf life (product resistance)
5. To increase line speed (productivity)

The acrylic acid based ethylene copolymers are the most cost effective of the adhesives available for polyethylene and aluminium foil. In coextrusion the recommended EAA coating weight is between 5-15 g/m^2. EAA copolymers with 5-10% by weight acrylic acid are most frequently used. If the acrylic acid level is increased much above 9% then delamination from the polyethylene may occur. The danger is that as the copolymer becomes increasingly acidic it loses its affinity with the non-polar polyethylene. It is important to establish if the weak link is between the foil and EAA, or between the EAA and polyethylene.

The data below compares adhesion (peel strength) between the EAA in coextrusion coating with LDPE on aluminium foil. The coating was applied on an OPET/foil laminate.

- EAA (6% AA): 6.5 N /15 mm from foil
- EAA (9% AA): 5.0 N/15 mm from LDPE

The higher acrylic acid content delaminates from the LDPE rather than the aluminium foil. The adhesion values are, however, high in both cases.

Zinc ionomers are also excellent as adhesive layers, but these are more costly than the EAAs and should only be used if improved product resistance can be demonstrated.

Figure 9.22 shows a typical 2-layer feedblock coextrusion set-up. In **Table 9.14** the adhesion values of a monolayer LDPE and a coextruded EAA/LDPE coated on an aluminium foil laminate are compared. The foil side of the laminate was corona treated in-line. The 0.7 N/cm adhesion value with LDPE at 310 °C is too low for commercial use. The coextrusion with EAA was inseparable from the aluminium foil surface.

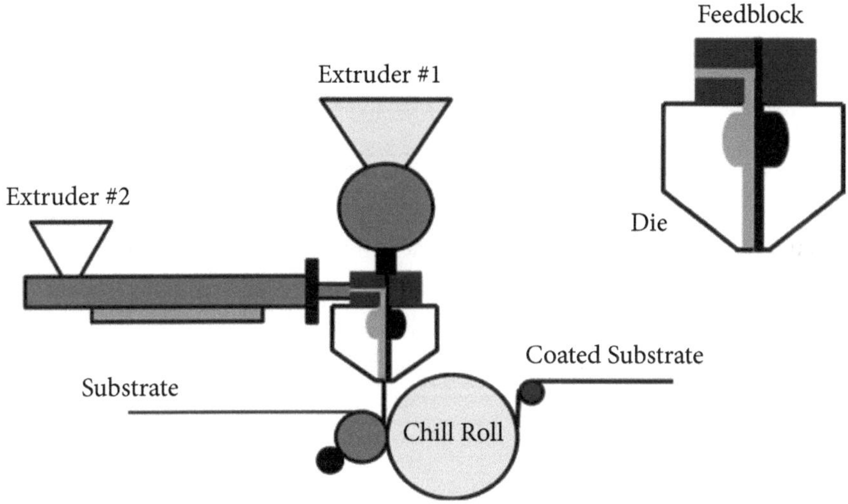

Figure 9.22

Simple 2-layer extrusion coating assembly

Table 9.14

Coextrusion compared to monoextrusion

Conditions	Mono	Coextrusion	
	Extruder #1 (LDPE)	Extruder #1 (LDPE)	Extruder #2 (EAA)
Melt temp. (°C)	310	291	287
Screw speed (rpm)	114	57	160
Pressure (bar)	63	55	76
Line speed (mpm)	150		
Coating weight (g/m^2)	30	15	15
Adhesion (N/15 mm)	0.7	Not separable	

In this instance coextrusion allows superior adhesion at conditions of minimum oxidation (low temperature). A significant reduction in smoke and odour will result from reduced temperature.

The enhanced adhesion provided by copolymers can enable significant increases in productivity. In this next example coextrusion was used to demonstrate an increase in line speed from 180 to 300 mpm. The application was a very fine highly sized paper coated with 30 g/m^2 of polyethylene. Maximum line speed for good adhesion with LDPE was **180** mpm at 325 °C. The monolayer LDPE coating was replaced by a 2-layer coextrusion using an ethylene copolymer as tie-layer.

EAA (9% AA) and EMA (6.5% MA) were compared as adhesives in this evaluation. Adhesive coating weight was 5 g/m^2 combined with 25 g/m^2 of polyethylene. The adhesion results at **300** mpm are shown in **Figure 9.23**. The LDPE was coated at 325 °C, the EMA at 310 °C and the EAA at 300 °C.

The EAA clearly enables the required line speed of 300 mpm to be easily achieved. LDPE and EMA adhesion fall below minimum specification at the same line speed. This paper substrate is highly compatible with the acid copolymer. These compelling results demonstrate how coextrusion can enable significant increases in line speeds: above 50% in this example.

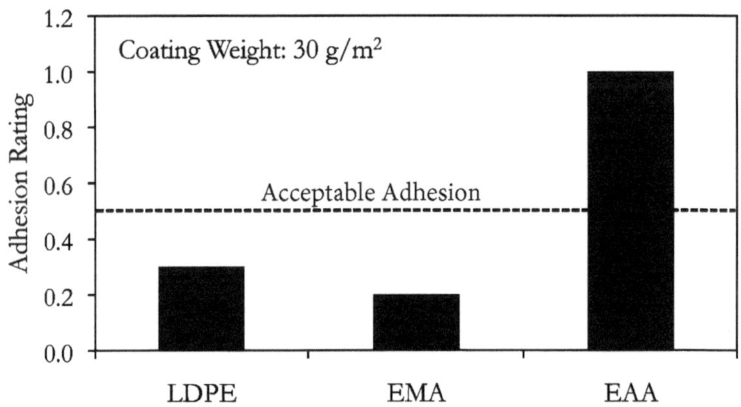

Figure 9.23

Adhesion of LDPE in comparison with copolymers (line speed: 300 mpm)

In the next experiment two EAA copolymers, one ionomer and a terpolymer were compared for their product resistance as extrudable adhesives. These were coextrusion coated with LDPE on an OPET/aluminium foil laminate. The final structure was as follows:

OPET 12 µm
PU adhesive 3 g/m^2
Al foil 8 µm
Coextrusion tie resin 10 g/m^2
LDPE 40 g/m^2

The adhesive layers are characterized in **Table 9.15**.

Table 9.15

Extrudable adhesives

Product	Melt index (dg/min.)	Comonomer
EAA	8.0	6% acrylic acid
EAA	8.0	9% acrylic acid
Ionomer	2.9	Zinc cation
Terpolymer	6.1	6% AA + methyl acrylate

A coextrusion coating line with the basic set-up similar to **Figure 9.22** was used. The extrusion temperature for both layers was standardised at 285 °C and the line-speed was 100 mpm. The air gap was 150 mm with in-line corona treatment of the aluminium foil. The final structure was converted into pouches on a vertical form, fill and seal machine and filled with a range of consumer products. Adhesion values after one month of storage at 40 °C are shown in **Figure 9.24**. The coatings have good resistance to sunflower oil, hair gel, cloves and lemon juice. Mustard

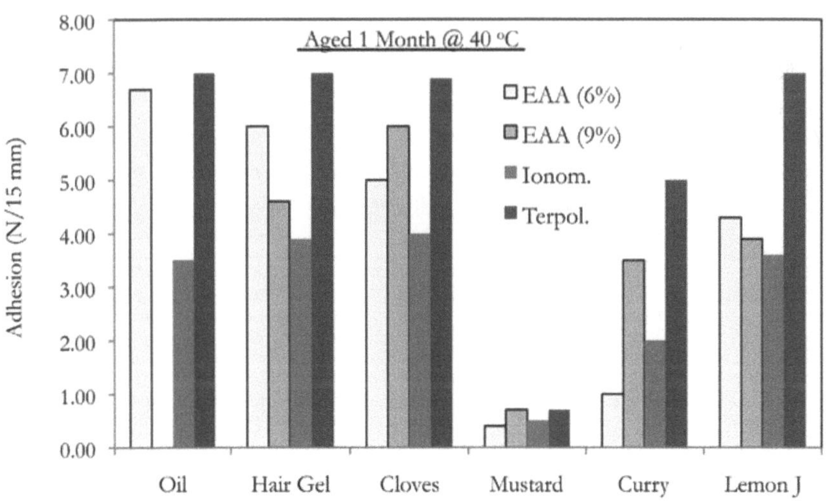

Figure 9.24

Product resistance after 1 month conditioning

appears to be the most difficult to contain. A higher temperature of 290-295 °C would have been preferred for the EAA and ionomer experiments.

The effect of reduced extrusion temperature on odour is shown in **Table 9.16**. Subjective judgements by an expert panel showed significantly reduced odour for the coextrusion versus the monolayer LDPE coating.

Table 9.16

Extrusion coating at 2 melt temperatures

Product	Temp.	Odour ranking
LDPE (monolayer)	310 °C	2.5 (high)
Coextrusion EAA/LDPE	290 °C	0.9 (low)

9.12 Some coextruded structures

In **Figure 9.25** a few of the layer combinations that can be produced with coextrusion equipment are illustrated. These structures were developed for cast film and sheet extrusion, but some are applicable to the extrusion coating process. The key advantage of the extrusion coating process is that additional layers can be added by coating and laminating.

The attributes that can be incorporated are as follows:

1. Improved heat seal by placing an ionomer or copolymer layer on the outside
2. An adhesive layer with improved compatibility to the substrate will enable higher line speeds and give improved product resistance in extrusion coating
3. Incorporating EVOH or another barrier polymer in the structure as an alternative to aluminium foil
4. Adhesive layers can very often allow significant extrusion temperature reduction resulting in reduced odour and stronger seals
5. Surface properties can be tailored by placing an outer layer of polymer containing additives or using a highly crystalline polymer such as HDPE. Chill roll sticking can be eliminated in this way
6. Bury scrap material in a core layer
7. Polymers that are prone to drawdown instabilities, typically LLDPE and HDPE can be carried on LDPE with its excellent balance of drawdown and neck-in
8. Degradable or corrosive polymers can be encapsulated in an inert polyolefin to avoid contact with metal surfaces inside the die

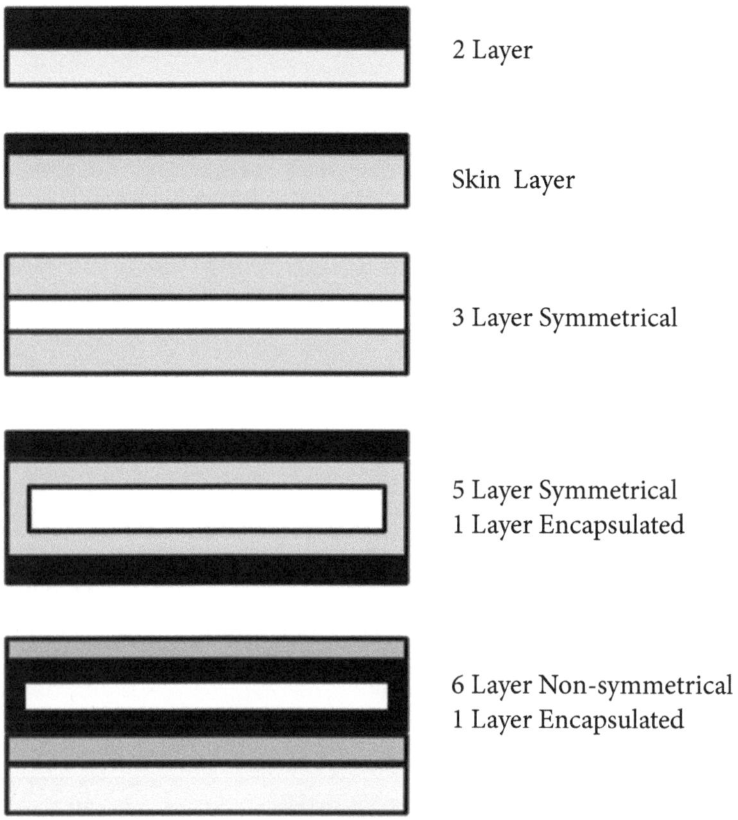

Figure 9.25

Some examples of multilayer structures

The development of encapsulation techniques makes it possible to isolate corrosive and heat sensitive polymers from direct contact with the die walls. PVDC has been successfully coextruded with die designs of this type. EVA (~26% VA) copolymers have good adhesion to PVDC and can be recommended as the encapsulation layer. Encapsulation is also a useful method of stopping expensive polymers from flowing to the outer edges of the die and ending up as edge trim in the waste material. This technique is used to reduce wastage of costly EVOH. The number of possible configurations is limited only by the imagination of the machine designer and processor.

9.13 Tubular film coextrusion

Coextrusion using the tubular film process is more complex than the combining feedblock. **Figure 9.26** shows a number of possibilities of blown film coextrusions

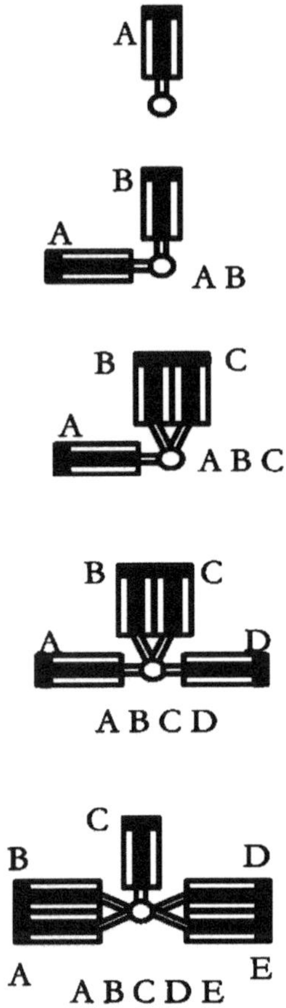

STRUCTURE	MATERIALS
A	ALL PEs
AB or ABA	A=PE etc. B=PE etc.
ABC or ABCBA	A=PE etc. B=PE or Tie resin C= PE, PA, EVOH
ABCBD or ABCBCBD	A=PE, PA B= Tie resin C=PE, PA,EVOH D=PE, PA
ABCDE or ABCBDBE or ABCDCBE	A & E= PE, PA B & D= Tie resin C= PE, PA, EVOH A & E= PE, PA B= Tie resin C= PE, PA, EVOH D= PE, PA, EVOH

Figure 9.26

Tubular film coextrusion layouts

with different arrangements of extruders and dies. The melt streams are preferably led from a port block into the die, which feeds the individual melt streams from the extruders to their corresponding flow channels. These are a series of concentric spiral cut mandrels, which distribute the melt streams independently. These overlapping layers are combined around a central mandrel just before the exit. One advantage is that the layer rearrangement phenomenon discussed in section 9.3 does not occur with annular dies.

However, coextrusion tubular dies can be very cumbersome and are difficult to calibrate, purge and maintain. Unlike the simple combining feedblock, these are dies within dies and become excessively large and heavy as the diameter increases from the inner mandrel with each addition to the outermost layer. This increase in diameter exposes the melt to a greater surface area, which increases pressure (shear stress) and creates the potential for interfacial instabilities and degradation (see **Figure 9.9**). The problem is multiplied by the number of layers. **Figure 9.27** shows a typical 5-layer design. This shows each layer as virtually independent dies within the whole. Thickness control must be adjusted independently for each layer.

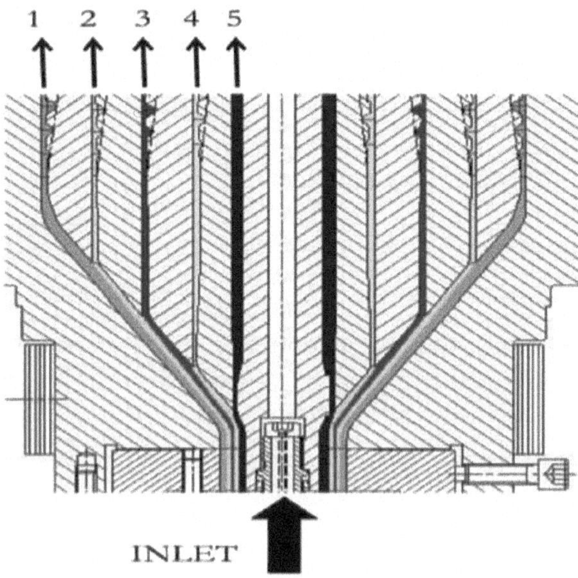

Figure 9.27

5-layer tubular coextrusion die

To avoid gauge bands on the finished rolls rotating or oscillating dies are used. A typical rotating die cycle is 7-8 minutes. These have a number of disadvantages in coextrusion:

- Increased susceptibility to leaks at the seals of the rotating mechanism
- Stripping and reassembly for maintenance is time-consuming
- Purge times can be excessive
- Die height increased by 1 m

Fixed dies with rotating haul-offs are preferred. These will have 360° oscillation and are equipped with turning bar systems that guide the layflat down to the winder placed adjacent to the extruders. The rotating time for the full cycle is around 20 minutes. **Table 9.17** compares subjectively 5-layer tubular coextrusion plants with rotating and stationary dies.

Table 9.17

Advantages of fixed die in 5-layer coextrusion

IBC: Internal bubble cooling	Die rotation – no IBC	Fixed die with IBC
Output	100%	145%
Purging time	100%	40-50%
Production with EVOH	100%	200%
Cost	100%	84%
Centring	Not possible	Possible from outside
Stripping etc.	0	++
Gauge variations	8-10%	4-6%
Melt channel length	100%	60%
Maintenance	100%	80%

The development of zigzag waves and interfacial instabilities is more prevalent with annular dies due to the higher pressures developed with low melt flow polymers and lower temperatures required for melt stability and to reduce demand from the bubble cooling unit. Flat die processes, as discussed earlier, run at higher temperatures with 'easier' flow resins and have virtually no cooling limitations.

Stackable die designs or pancake dies were developed to reduce height. The spiral section is machined in a flat horizontal surface or 'pancakes', which are stacked on top of each other to minimize space. The melt distribution occurs in a plane that is perpendicular to the axis rather than in the usual concentric flow direction. The melt streams are fed to each layer from the side of the mandrel. One disadvantage is that the melt is internally split several times as it passes through a number of channels towards the centre of the die. This excessive melt splitting, together with limits in the number of spirals and excessive increases in diameter of the outer layers, require a disproportionate number of bolts to seal each mandrel from melt leaks. Modifications such as tapered cones have been developed to minimize leaks. However, surface area is significantly increased limiting the size of die that can be run successfully.

In the writer's opinion where more than 3 layers are required the much simpler combining feedblock system with flat dies should be considered. Flat dies have the advantage of diverters and flexible lips, which simplify fine tuning of layer thickness distribution. The absence of high velocity cooling air eliminates the effect of turbulence on film thickness control and chill roll cooling enables much higher line speeds.

Chapter 10

Coextrusion tie-resins

In coextrusion interlayer adhesion is critical. In many applications the polymers used will adhere to each other usually by intermolecular diffusion forming very strong bonds. Typical examples are the polyolefin homopolymers and copolymers. With the exception of polypropylene these polymers are compatible with each other. Where the polymers are incompatible an additional layer is required to form a bonding bridge between adjacent layers. In the case of coextrusion of styrenics with a polyolefin, a copolymer such as EVA (26-28% VA) will act as a useful tie-resin. EVA and EMA copolymers will also work with PVDC. In the case of totally incompatible polymers, such as a polyolefin with EVOH or polyamide (PA) specialized adhesive resins are required.

The key attributes of the principal ethylene copolymers and ionomers are summarized in **Table 10.1**. The characteristics of the acid and ester copolymers in terms of adhesion are differentiated.

Table 10.1
Attributes of ethylene copolymers

Material	EMA*	Ionomer	EAA	EVA
Processing range (°C)	150-320	165-305	175-300	230-260
Thermal stability	Excellent	Good	Good	Fair
Adhesion to:				
BOPP/PP	Good	Poor	Poor	Fair
PAPER	Good	Excellent	Excellent	Good
PET/PETG	v. Good	Poor	Poor	v. Good
PVC (rigid)	Good	Poor	Poor	Fair
Al foil/metallization	Poor	Excellent	Excellent	Fair
** Applies to other alkyl acrylates*				

The compatibility and adhesive properties of ethylene homopolymers and copolymers can be enhanced by grafting a functional group onto the polymeric chain. In **Figure 10.1** a copolymer is compared schematically to a grafted polymer. In the former case the comonomer is incorporated into the growing polymer chain with the principal monomer. In the grafted case the comonomer is placed onto the fully formed polymer molecules. These grafts form pendent functional groups that can achieve excellent adhesion and improve blending dispersion at low concentrations of the grafted monomer.

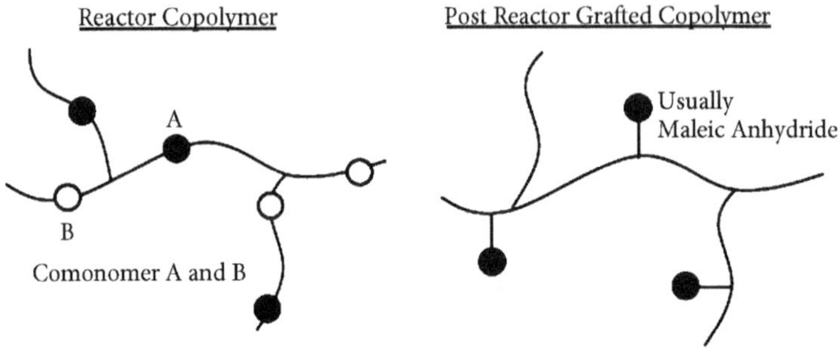

Figure 10.1

Copolymer and grafted polymer

Figure 10.2 shows the simplified reaction. The peroxide initiator forms a free radical (R•) by heating and activates the polymer by hydrogen abstraction. Chain cleavage occurs to form free radicals, which enable the grafting of the maleic anhydride onto the polymer backbone to take place. The skill in this technology is to avoid crosslinking or drastic chain scission of the polymer. A monomer which does not easily homo-polymerize is preferred. Maleic anhydride (MAH) is normally used in these reactions. This monomer is non-corrosive and reacts favourably with most polyolefins in the presence of peroxide initiators. The reaction is carried out in twin screw extruders designed with appropriate monomer and initiator ports. An inert gas (nitrogen) should be used to purge the machine during processing. The screws must be designed to optimize distributive mixing to ensure random grafting of the anhydride. The layout is shown in **Figure 10.3** for a co-rotating screw machine. The finished material must

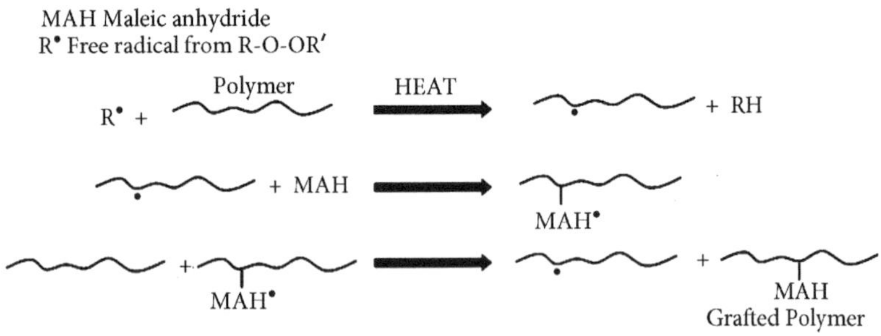

Figure 10.2

Grafting of functional groups on polyolefin backbone

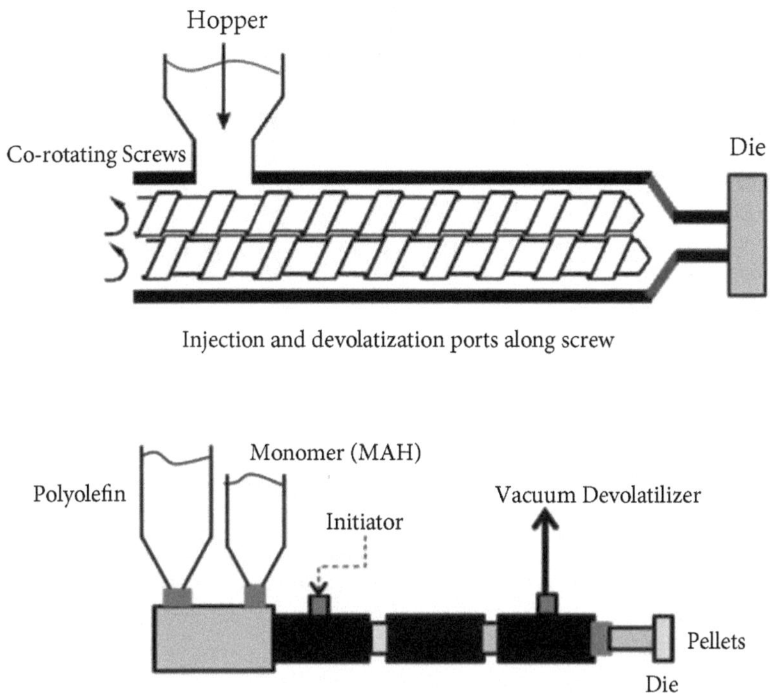

Figure 10.3

Twin screw extruder reactor

be thoroughly degassed to remove any residual trace of monomer. The grafted MAH content is usually <0.5% by weight.

An important advantage of grafting is that the properties of the backbone polymer remain unchanged. In copolymerization crystallinity is invariably reduced, which affects the polymers' properties.

The functionality of the MAH grafted polyolefin is shown schematically in **Figure 10.4**. Attractive bonds are formed between the anhydride with either the hydroxyl (-OH) groups in EVOH or the amine groups in polyamides. The anhydride is believed to be more reactive with amines than hydroxyls. The backbone polymer chain intermixes with the polyolefin layer by diffusion.

Commercial products are very often made by producing this material as a concentrate, which is let down in other polymers. LLDPE and HDPE are the easiest of the polyethylenes to graft with maleic anhydride and polypropylene the most difficult. LLDPE disperses easily in other polyolefins and is a useful host resin. The dilution level of the maleinized concentrate will affect adhesion.

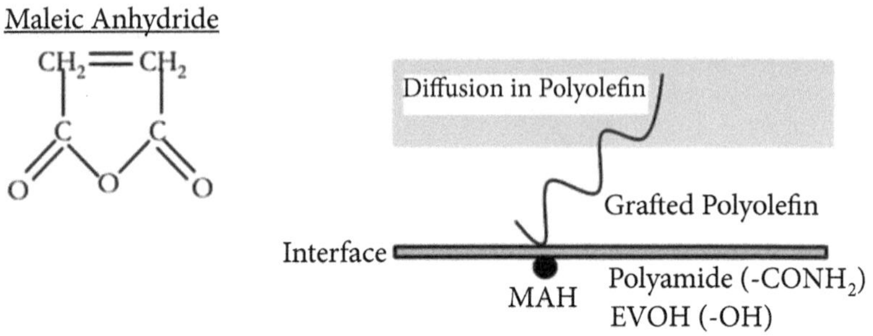

Figure10.4

Functionality of grafted polyolefin

For steam sterilizable applications HDPE is used. HDPE based systems will also offer improved resistance to lubricating oils and hydrocarbon fuels.

In some cases an ethylene propylene elastomer (EPDM) is added to improve dispersion and toughen the product. The presence of a well-dispersed rubber phase reduces crack propagation at the interface, which increases bond strength. The dispersion of EPDM in the continuous phase of the host polymer is significantly enhanced by grafting maleic anhydride through its double bond. The quality of the dispersion can have a profound effect on bond strength.

Polyisobutene (PIB) is also quoted as a rubber modifier in some of the literature. However, dispersion of PIB in polyolefins is less effective than ethylene propylene rubbers. EPDM grades with high ethylene content and low Mooney viscosity (low MW) should be used. These can form a very finely dispersed phase in the continuous phase of the host polyolefin.

Polypropylene can be blended with MAH-grafted LLDPE using 2-5% EPDM as compatibilizer. Such blends can enable the coextrusion of PP with other polymers. This approach is simpler than trying to graft directly onto the PP backbone.

These highly specialized materials have also been finding increasing use as compatibilizer to blend waste plastics for recycling purposes. Useful blends of recycled EVOH or polyamide with polyethylenes can be made by the use of ~5% of MAH-grafted polyolefin as compatibilizer. These can be buried in core coextruded layers.

These chemically modified polymers are formulated according to the demands of the application. Some examples are in **Table 10.2**. The LLDPE and HDPE grades may contain 2-5% rubber. To minimize cost these resins can be diluted by the processor with LD/LLDPE. The bond strength will vary with dilution. As can be seen in **Table 10.2** the maleic anhydride weight percent concentrations are low.

Table 10.2
Commercial MAH grafted polymers

MI (dg/min.)	Composition	VA (%)	MAH (wt.%)
1.7	EVA/LLDPE (85/15%)	3.9	0.10
2.0	EVA/LLDPE (80/20%)	3.7	0.12
3.2	EVA/LLDPE (80/20%)	7.5	0.13
2.0	LD/LLDPE (60/40%)		0.31
1.5	LDPE	-	0.27
1.1	LLDPE	-	0.36
3.2	LLDPE	-	0.30
3.5	LLDPE	EPDM mod.	0.24
0.85	HDPE	Rubber mod.	0.25

Tables 10.3 and **10.4** compare adhesion values of blown and cast films with PA 6 (polyamide) and EVOH. The samples were based on 5-layer coextrusions with LD/LLDPE blends as the support polymer. The results show significant differences between the two film processes. The differences between the processes are more marked with the polyamide (PA6) samples than the EVOH. Quenching is very rapid in the cast film process, which has a profound effect on crystallization and may affect the interaction at the interfaces.

Table 10.3
Adhesion of polyethylene/polyamide 6

MAH-grafted layer	MAH (wt.%)	Bond strength (N/15 mm)	
		Blown film	Cast film
*LLDPE rubber mod.	0.24	10.0	1.5
LLDPE	0.36	Inseparable	0.75
EVA/LLDPE 80/20	0.13	0.96	Inseparable
EVA (8% VA)	0.09	0.2	0.17
EVA/LLDPE 80/20	0.12	0.72	0.55

**Diluted to 10%.*

Table 10.4

Adhesion of polyethylene/EVOH

MAH-grafted layer	MAH (wt.%)	Bond strength (N/15 mm)	
		Blown film	Cast film
LD/LLDPE (60/40%)	0.31	9.70	7.22
LDPE	0.27	8.06	2.25
LLDPE/EPDM mod.	0.30	0.18	0.49

Chapter 11

Substrates and films

To the consumer the complex construction of packaging materials is barely noticeable. As an example a simple snack food pouch may include 10 layers in its construction.

An in-depth understanding of the synergy between the polymer, the substrate and the process capability is essential if the extrusion coater and laminator is to manufacture products at an optimum cost and performance balance. This chapter presents an overview of the key substrates and films available to the fabricator and converter.

The handle and extensibility of a packaging film will depend on its stiffness and thickness. Generally a film or laminate should have high stiffness and minimum extensibility, i.e. no stretch. **Table 11.1** shows the tensile moduli of a number of packaging materials. The EVOH values are based on blown film.

Table 11.1

Tensile modulus of films

Film	MPa	kpsi
Al foil	6900	10000
OPET	3450-3800	500-550
PA 6 cast film	620-560	90-125
PA6 mono-oriented	1380-1520	200-220
PA 66 cast film	620-830	90-120
PA 66 mono-oriented	2070-2420	300-350
EVOH 32 mol% C_2	3040	440
EVOH 44 mol% C_2	1450	210
BOPP	1590	230-270
LLDPE blown film	170-280	24-40
LDPE blown film	170-210	25-30
HDPE	500-1400	72-203
EVA	50-200	7-30
EAA blown film	110-160	16-23
Ionomer blown film	140-410	20-60
Polystyrene (GP)	2760-4140	400-600
Polycarbonate (PC)	4125	350

11.1 Adhesion of cellulose/paper materials to PE

The adhesion of polyethylene to cellulose materials is induced via the carbonyl groups (-C=O) generated through oxidation. However, the presence of certain sizing agents, water-repellent chemicals and fillers in the pulp can reduce adhesion even when the polymer is heavily oxidized.

The work of Zisman[1] in the 1950s on the wetting of polymer surfaces established that adhesion to paper could be predicted by the concept of critical surface tension. Papers that have critical surface tensions below 31-32 mN/m had poor adhesion to polyethylene compared with those having higher surface tensions.

Table 11.2 shows some critical surface tension values for cellulose derived films and papers. The broad range in surface tensions within each category shows the importance of evaluating each product individually if variations in adhesion are to be avoided in the extrusion coating operation.

The adhesion of PE to paper and board is also improved by increasing paper surface roughness. This is achieved by different degrees of calendaring. Adhesion of LDPE increases with surface roughness to a critical coating weight above 15 g/m^2.

Table 11.2

Critical surface tension of cellulose materials

Product	**mN/m**
Regenerated cellulose film (RCF)	42
Paperboard (cardboard)	31-48
Glassine paper	35-36
Acid-made paper	38.7
Alkaline-made paper	33.5
Acid-and-alkaline-made paper	24-43
Unsized cellulose film	39-52

11.2 Paper and paperboard

Paper is the generic term used to describe matted webs usually made from wood fibres. The pulp is deposited from a water suspension onto a screen to form a substrate. The excess water is removed and the substrate is converted into a sheet by pressing, drying and calendering.

1 Zisman, W.A., Contact Angle, Wettability and Adhesion, ACS Advances in Chemistry series, **1964**, 43, p. 1-51

Paper properties will depend on the characteristics of the wood pulps used in the manufacturing process. Pulps are made from softwoods (conifer) or hardwoods (deciduous). Mechanical or chemical processes or a combination of the two are used to prepare the pulps. The pulps can be left unbleached or are bleached to varying degrees by a number of techniques.

In view of the wide variation in density that is possible with current technology the demarcation line between paper and paperboard (cardboard) is blurred. Structures **less** than 305 μm (0.012 inch) thick are usually defined as paper regardless of their weight. ISO standards state that papers above 250 g/m^2 in weight will be described as paperboard or cardboard. There are exceptions however, such as drawing and blotting papers whose thickness may exceed 300 μm and still be classified as paper.

Papers are classified as either coarse or fine. The coarse papers are unbleached and usually made from softwood pulps. Fine papers are produced from bleached pulps and used in applications requiring fine graphics or art work, and characteristics such as a barrier to fluids and grease.

The structure of paperboard is shown in **Figure 11.1**. The substrate is made up of different plies of homogeneous composition. The topside or liner is the better quality surface of the web and is very often white. Double lined boards will have high quality finishes on both sides. The underliner is the ply of fibres between the topside and the middle. Backs or backs liner is the outside ply as shown in **Figure 11.1**.

The term **furnish** is used in paper technology to describe the constitution of the various materials that are blended in the aqueous pulp suspension from which the different plies or layers are manufactured. Typical constituents are fibrous materials, fillers, sizing agents, pigments, optical whiteners and dyes.

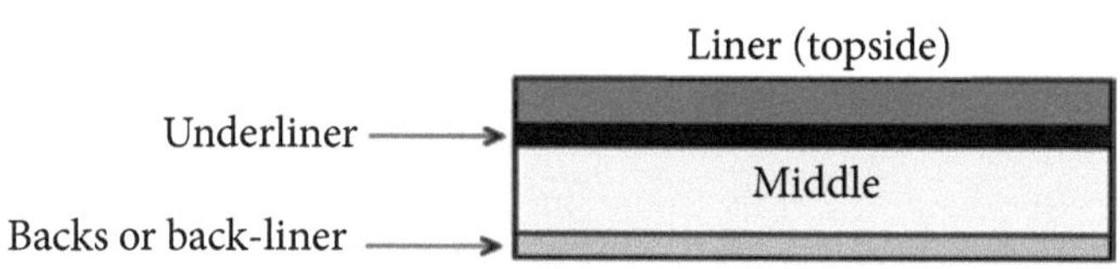

Figure 11.1

Typical multiply paperboard

A number of papers and paperboards used by the extrusion coater are briefly described below.

Duplex. Board constructed from two layers of different quality or furnish.

Triplex. Board made up of three layers of different furnish. The external layers may be from the same composition of fibres.

Multiplex. As the name implies, these are paperboards made from more than three layers of different composition. The external layers may be equivalent. The liner or top side is made of good quality fibres. Multiplex is sometimes referred to as multilayer board.

Kraft papers. High strength kraft paper is made from no less than 80% sulphate wood pulp. The brown paper is finished by calendering or machine glazed. The surface is normally acceptable for printing. Kraft papers are used for a variety of wrapping applications, grocery bags, multiwall sacks, envelopes, and speciality packaging applications requiring strength.

Kraft papers are extrusion coated with polyethylene to provide moisture-proofing for multiwall sacks and bags.

Bleached papers. Bleached papers are used in packaging applications where the priority is decorative rather than strength. The soft, white pulps must be receptive to the chemicals required for functional properties. Whiteness and brightness will enhance print quality and generate a perception of cleanliness and quality.

Surface smoothness and whiteness can be further improved by clay coating. Clay coated papers are produced by the application of aqueous mixtures of clay and other mineral pigments with organic binders. These aqueous coated papers are used in applications where aesthetic appeal and presentation are important.

Glassine papers. The hydrophilic nature of cellulose makes it a good substrate to resist penetration by hydrophobic liquids and greases. By mechanical refining, a pore-free consolidation of the substrate is achieved. Refining fibrillates, breaks and swells the cellulose fibres and allows the interstitial spaces to be filled, thus drastically reducing its porosity and enhancing resistance to grease penetration.

Glassine paper is manufactured by super calendering. The process is carried out at high temperature with steam and nip rolls of different hardness to achieve a fine polish. The paper fibres are flattened out facing the same direction, which generates very intimate interfibre hydrogen bonding to minimize liquid penetration. Greaseproof paper is made from a glassine base paper.

Glassine can be extrusion coated with LDPE to add sealability and reduce moisture transmission rate (MVTR). High temperatures and corona treatment are required to bond the polymer to the smooth surface.

Photographic paper is discussed in section 2.3.

11.3 Aluminium foil

There are two methods of casting aluminium to a strip. The first method is to cast an ingot, which is hot rolled to a thick strip and the second is to cast a 2-6 mm strip directly by continuous casting. Both types are then cold rolled to light gauge foils. Cold rolling is carried out in 6-8 steps reducing thickness by 50% in each step. In the final step the foil is doubled into two layers and sprayed with rolling oil to facilitate separation. After separation the outer side is bright and the inner side will have a dull appearance. For flexible packaging applications the foil is annealed at 300 °C to recrystallize the grain structure and evaporate the residual rolling oils. The fresh aluminium surfaces will immediately react with oxygen to form a well adhered layer of oxide. This amphoteric oxide (Al_XO_Y) layer is critical as it is receptive to the very hot oxidized (-C=O) LDPE delivered from the extrusion coating die. The very hot polymer should lock with the oxide layer whose porosity and strength will depend on the manufacturing process. Surface treatment, i.e., flame or corona will enhance reactivity.

In extrusion coating with EAA copolymer (acrylic acid) the aluminium oxides will react as bases with the acid thus forming much stronger bonds than possible with LDPE. Zinc ionomers, which are partially neutralized acid copolymers, interact through a similar mechanism.

It is important that during metal casting the molten aluminium is filtered to remove the finest particles of ceramic material or other contaminants in the process. During these casting processes and thermal treatments primary and secondary phases are formed, which are distributed in the structure. These phases usually consist of compounds of aluminium, iron and silicon. The quality of the foil for the converter is defined by the homogeneity of its grain structure and distribution and size of these phases. To maximize resistance to acids the iron content must be reduced to an absolute minimum.

Classifications of some typical aluminium foil products are shown in **Table 11.3**.

Table 11.3

Types of aluminium foil

Designation	Composition	Thickness	Properties/use
AA 1050A	Al>99.5%	6-9 μm	Supported on paper/plastic
AA 1200	Fe 0.5-0.7%	>7 μm	Wrapper, high strength
AA8079	Fe 1.1-1.3%	>12 μm	Very ductile
AA 8014/8006	Fe>1% Mn~0.3%	>10 μm	Highest strength

The transport of small molecules through plastic films is mainly by diffusion. Diffusion rate is controlled by the composition, morphology and thickness of the polymer. In contrast transport through aluminium foil is by flow through holes or cracks. When free from such defects aluminium foil is a perfect barrier. However, a typical 9 μm foil can contain anything from zero to 1000 pinholes per square metre. Its barrier performance will, therefore, depend on the size and incidence of pinholes and cracks in the foil.

Some data comparing the incidence of pinholes in commercial grade aluminium foils are shown in **Table 11.4.**

Table 11.4

Incidence of pinholes in commercial aluminium foils

Foil thickness (μm)	Typical pinholes/m^2
7	300-500
9	30-40
12	approx.10

The combined pinhole area even in a highly porous foil is not likely to amount to more than 0.001% of the total area. In a typical extrusion coated laminate, the aluminium will be sandwiched in between two layers of LDPE or a similar polymer. The transmission of gas or moisture will now depend on the permeability of 0.001% of LDPE surface covering the pinholes. It is easy to conclude that the ingress or loss of gases through this tiny surface area of pinholes covered with polymer will be too low to be measurable with existing equipment. A soup pouch will for instance have a surface area of 0.03 m^2. The total area of pinholes per pack is clearly extremely small so providing that these are covered by the polymer the laminate will be essentially impermeable. In effect the transmission of gases through the heat seal fins of the pouch will exceed any ingress or loss through its walls.

For the extrusion coater the main problem associated with the use of aluminium foil is to ensure good adhesion to the coating. The foil must have perfectly clean surfaces with excellent wettability. Any residual rolling oils on the surface will inhibit adhesion. A wettability test can be used with a solution of 5% ethanol in distilled water. The interaction of the aluminium surface is through the oxides embedded on its surface as discussed earlier.

For non-transparent applications, aluminium foil laminates are the best flexible barrier material available for packaging. Aluminium combined with paper and polyethylene provides an outstanding combination of attributes.

These are:

- Excellent barrier to gases, moisture and light
- Stiffness and strength from the paper or board
- Excellent machineability on high speed packaging lines
- Excellent sealability of LDPE, copolymer, ionomer etc.
- Good interlayer adhesion
- Heat seal by induction heating
- Easily printable web
- Dead-fold

The metallic layer rapidly conducts heat away from the hot sealing area thus enhancing throughput on form, fill and seal machines. Aluminium has a conductivity that is 600 times higher than a polyolefin.

Negative attributes include the following:

- Poor flex crack resistance
- Attack by acid foods (pH <6.4)
- Not microwaveable
- Recycling is difficult

Paper 'poisoning' can also occur with paper-foil laminates. This phenomenon inhibits the adhesion of the PE extrusion coating on the exposed foil side. Poisoning occurs when the foil side is tightly pressed against the paper backside in a reel and is thought to be due to off-setting of certain compounds from the paper to the foil surface. The best way of avoiding this problem is to extrusion laminate the foil to paper and coat in **one pass** on a tandem machine.

11.4 Flexible packaging films

The films that are commonly used for lamination and coating fall into two groupings.

1. Plastic Films: OPET, BOPP and polyamide (nylon)
2. Regenerated cellulose films (RCF)

RCF films have been substituted in many packaging applications by BOPP.

These films are available with or without barrier coatings (PVDC), metallization, lacquers and various modes of orientation. The films will very often be reverse side printed prior to lamination.

Adhesive lamination is the recommended process to combine these plastic films with polyethylene. When long runs are feasible extrusion coating can be used. The risk of stress relaxation caused by the hot melt (310-320 °C) may limit the choice of suitable films. BOPP films are more heat sensitive then OPET. A coextruded tie-layer or chemical primer will usually be required for acceptable adhesion. If gloss chill rolls are used film blocking and chill roll sticking may occur.

In general adhesive lamination is the preferred option as it permits a broader choice of LL/LDPE/EVA films. A wide range of isocyanate based adhesive formulations are available for most packaging requirements. The development of reactive solventless systems has eliminated the need for large drying ovens. **Table 11.5** summarises the permeability and grease resistance of the principal packaging films.

Table 11.5
Permeability and grease resistance of 25 µm films

Property	RCF (nitrocellulose coated)	LDPE	HDPE	Cast PP	BOPP	PVC	OPET	PA-6
OTR (cm^3/m^2.day.atm.)	670 (0% RH) (50% RH)	6500-8500	1600-2000	3700	2000-2500	150-350	40-50	30-40
MVTR (g/m^2.day @ 90% RH)	5-15	15-20	5	10-12	7	30-40	25-30	300
Oil/fat resistance	Excellent	Some swell	good	good	Excellent			

11.5 Regenerated cellulose film (RCF)

Cellulose is a long chain carbohydrate that includes a large number of hydroxyl groups per molecule.

These groups induce strong intermolecular attractions via hydrogen bonds. The material is **not** thermoplastic. The process for making RCF is based on chemically regenerating viscose, a dissolved cellulosic compound, into a thin cast film. Viscose is made by adding dilute sodium hydroxide solution to sodium cellulose xanthate. Xanthate is derived from soft wood pulps treated with alkali, depolymerized and reacted with carbon disulphide. A complex process then follows where the viscose solution is further depolymerized following a filtration sequence and separation of the excess sodium hydroxide. The viscose is pumped through a slot die and cast in a sulphuric acid bath where the cellulose sheet is regenerated. The cellulose then passes through a series of tanks to remove sodium sulphate and is blended with softeners and plasticizers. The film is finally dried to 7-8% moisture level, wound on steel cores and made ready for converting.

RCF plasticized with glycerol is usually recommended for extrusion coating. The films are coated for moisture protection (nitro-cellulose lacquer), heat seal and barrier (PVDC). Cellulose films have traditionally been used in a wide variety of packaging applications that include confectionery, tobacco, drugs, candy, cookies, nuts, and snack foods. However, over the past thirty years, BOPP has successfully replaced RCF in most of its key applications, owing to lower cost and a less polluting manufacturing process.

Table 11.6 shows the barrier properties and coefficient of friction values of some typical RCF films. Thickness ranges from 20-45 microns. One advantage that RCF films have over BOPP is dead fold, which enables twist wrapping of candy/sweets and other confectionery products.

Table 11.6

Properties of RCF films

Film	OTR (cm^3.µm/ m^2.day.atm.)	MVTR (g/m^2.day @ 38 °C)	Heat Seal (°C)	COF
Nitro-cellulose coated, non-heat seal	700-1000	140	-	2-3
PVDC coated, heat seal	185	150	75	2-3
Uncoated, non-heat seal	1500	4000	-	4-5
Nitro-cellulose coated, heat seal	700-1000	160	77	2-3
OTR @ 50% RH. MVTR @ 99% RH. COF/ASTM D1894.				

11.6 Oriented polypropylene film (BOPP)

BOPP can be manufactured via blown and cast film processes. In the tubular film process the polypropylene is extruded into a thick tube. The tube is cooled in a water cascade or bath and its internal surface cooled by proprietary devices, such as chilled mandrels. The cooled tube at around 30 °C is pulled through a second nip and reheated to ~160 °C. Air is injected and a second bubble is formed. The machine direction (MD) orientation is controlled by the speed differential between the second set of nip rolls and the haul-off equipment. The blow-up ratio determines the degree of transverse (TD) orientation. The film is then passed through an annealing chamber and heat set into a stabilized web. Well-balanced films are produced by this process.

The alternative process is based on first preparing a cast film and then biaxially stretching the film in a continuous process. Similar temperatures to the tubular film process are used to orient the film. The heated film is firstly stretched in the machine direction (MD) by a system of nip rolls. Stretching is achieved by increasing the velocity between the sets of nip rolls. Transverse direction stretching is induced by a tenter frame, which clamps the edges of the film and continuously pulls the film outwards. The clamped edges are trimmed away and recycled in the film extruder(s). An annealing stage to heat set the film follows in-line. Balanced orientation at high draw ratios is not easily achieved with the tenter technique. This is mainly due to the sequential nature of the process, whereas with the bubble process the biaxial stretching is carried out in a single step. State-of-the-art tenter machines are 10 m wide operating at line speeds of 500 mpm.

A new development in film orientation technology is the commercialization of simultaneous tenter stretching machines using linear motors. In this process simultaneous orientation is achieved by diverting the clamps in the MD and TD in a closed loop. Each clamp is driven by a linear motor. Each of these is controlled separately. The stretching is contact free and allows a high degree of flexibility in the orientation profile of the film. Competitive production speeds are claimed.

BOPP films have clarity and sparkle that rivals cellulose films plus the added advantage of excellent moisture resistance. The film extrusion stage of the process includes coextrusion options usually to combine at least three layers. These may include copolymers for improved heat seal and/or to enhance physical properties. However, special inks are required for printing. OPET and RCF films are easier to print than polypropylene, which must be intensely treated.

BOPP films are available with a variety of coatings to improve their heat sealability, gas barrier and appearance:

- Two-side acrylic coated BOPP of 18-25 μm thickness depending on application
- Acrylic coated/BOPP/PVDC films from 20-35 μm thickness
- PVDC/BOPP/PVDC films from 20-25 μm thickness

- Metallized BOPP - good barrier and metallic appearance
- Pearlized BOPP (opaque) films for confectionery

PVDC coated BOPP films have 25 μm OTR values below 20 cm^3/m^2.day. The acrylic coating, usually applied from solution, functions essentially as a heat seal layer and adds some stiffness. High barrier ceramic (silica) coated films are also commercially available. These offer barrier properties similar to metallization. Recent developments in coextrusion and orientation techniques have enabled the production of oriented 7-layer PPRC/PP/EVOH coextrusions with improved gas and moisture resistance. PPRC is propylene random copolymer containing 4-6% ethylene. Tie-layers are used to bond the EVOH to the polypropylene.

11.7 Oriented polyester film (OPET)

Biaxially oriented polyester film (OPET) is used in a wide range of applications including magnetic tapes, pressure sensitive tapes, solar panels, labels, heat lamination and thermal insulation. Usage in packaging accounts for around 20% of production. OPET is normally produced from a cast polyethylene terephthalate film that is biaxially oriented in a tenter frame. For packaging use the film is oriented to the same degree in both directions. This imparts excellent balanced physical properties. The orientation process improves tensile strength, barrier, clarity and stiffness. OPET films have very good dimensional stability. A range of grades with corona or chemical treatment are available to facilitate converting.

For the converter, OPET film has many attractive properties:

- Dual ovenable up to 220 °C
- Resists temperatures down to -40 °C
- Machineability: non-extensible
- High tensile strength
- Dimensional stability over a wide temperature range
- Excellent transparency
- Excellent printability: surface tension 43 mN/m
- Outstanding chemical resistance
- Low scalping

OPET films are supplied as straight grades, coextruded, metallized, oxide (silica) coated and/or PVDC coated for sealability and improved barrier. In flexible packaging 12 µm thick films are normally used. The high film cost is, therefore, partially off-set by its high yield. The material forms an excellent base for lidding stock especially for ovenable trays. In packaging applications OPET films are available with various heat seal layers. EVA or PVDC sealant layers are normally used.

There is a growing market for heat-activated films on substrates such as prints, book covers, photographs, maps, etc. These films are usually made by extrusion coating low melting point EVA (16-18% VA) on OPET film. A typical laminate is 12 µm OPET coated with 20 µm EVA. The EVA layer will easily melt under heat and pressure and adhere to the substrate.

11.8 Polyamide film (nylon)

Polyamides (PA) are tough materials with high tensile strength and high melting point. Films extruded from polyamide have good oxygen barrier, but relatively high water vapour permeability. Polyamide films in the form of laminations or coextrusions with polyolefins are used in vacuum packaging for coffee, meat and cheese packs and boil-in-bag applications. Mono-oriented and biaxially oriented polyamide films are also available for lamination and coating. These films have superior strength and barrier compared to the unoriented films.

The key attributes of PA based films are:

- Excellent thermoformability
- Excellent flex-crack resistance
- Abrasion resistance
- Excellent aroma and odour barrier
- Moderate gas barrier
- Outstanding grease resistance
- Strength/toughness

Polyamide films can be extrusion coated with polyethylene with an aqueous **imine** primer. High temperatures in the range of 315-320 °C are used. Bond strengths of 3-4 N/15 mm width of specimen are obtained at coating line speeds of 200-300 mpm. Several hours ageing is required to maximize the bond strength. Coating weights range from 15-30 g/m^2.

Coextrusion is the more usual process to combine PA with polyethylenes. The MAH grafted PEs are used as tie-layers although ionomers have also been successfully coextruded with PA. The polyamide 6 grades (polycaprolactam) are preferred. Both blown and cast film coextrusion processes are used.

PA films laminated to a PE seal layer form an excellent base for the bottom web of thermoformed packages used for processed meat, cheese and fish. When oriented the properties of PA films are further enhanced as shown in **Table 11.7.** Although OPA has a higher OTR than EVOH its cost can be 40% lower thus enabling uses in less demanding applications.

BOPA is more costly to produce than either BOPP or OPET. However, the film has outstanding strength and is used in very demanding applications.

A useful attribute of PA is that it can be coextruded with EVOH without the need for a tie-layer. A PA/EVOH/PA coextrusion has excellent barrier and aroma retention. The coextrusion can be laminated or coated with polyolefins to add sealability and moisture protection. Flex crack resistance of these films is outstanding.

Table 11.7
Properties of polyamide films

Property	Units	PA-6, non-oriented	PA-6, mono-oriented	PA-6, biax. oriented
Tensile strength MD/TD	MPa	83/69	345/69	221/221
Elongation MD/TD	%	400/500	60/450	90/90
Tensile modulus MD/TD	MPa	690/793	2069/690	1724/1724
Initial tear strength MD	N/mm	193	251	77
Tear propagation MD/TD	N/mm	13.5/13.5	15/39	4/4
OTR $cm^3.\mu m/m^2.day$		10.1	9.3	4.7
MVTR $g/m^2.day$		7.1	7.1	6.7

The markets for these films are:

- Barrier laminates for pouches in horizontal and vertical form, fill and seal machines
- Lidding structures
- Bag-in-box food service items
- Typical sealing layers include LL/LDPE, EVA and PVDC

11.9 HDPE coextrusions

High density polyethylene (HDPE) film is opaque and crisp. The low MVTR of HDPE provides good protection from humidity and will minimize moisture loss from the packed product. Metallized films can be used to further enhance moisture resistance if required. HDPE films coextruded with EVA and LLDPE have been replacing glassine in many markets such as breakfast foods and pasta products. A typical coextruded structure is HDPE (40 μm)/EVA (8 μm), although improved heat seal properties are claimed by replacing the EVA with metallocene-catalyzed LLDPE (mLLDPE). For maximum moisture resistance the highest density HDPE grade should be used.

Another option is a 3-layer coextrusion of HDPE/HDPE/Ionomer. A peelable seal is formed by blending zinc and sodium (50/50) ionomers. Different grades of HDPE can be used in each layer.

HDPE is easily blended with other polyethylenes to control properties and minimize costs. The consumer bag and bin liner markets use a variety of blends with recyclates to reduce cost.

11.10 Metallized films

Metallized films are made by condensing a vaporized metal on the film surface. The operation is carried out in a vacuum chamber at around 3.1 kPa pressure. Aluminium is normally used for the production of metallized films for packaging applications. The amount of aluminium coated on the film is measured by optical density, light transmission or electrical resistance. Optical density is the most widely used specification method. Coating thickness is in the region of 0.05 μm.

In packaging, metallized films are used to improve barrier to gases, moisture and UV light. The sparkling metallic appearance of the film is equated by the consumer with freshness and quality. Metallized films are used mainly to improve the performance of films such as OPET and BOPP. The improvements in oxygen and moisture barrier resulting from metallization of three plastic films are shown in **Table 11.8**. The BOPP is the poorest in oxygen resistance.

Metallized films are coated or laminated with heat sealable polymers or lacquers for use in packaging. The metallized film is also laminated with reverse printed clear film to provide improved durability and depth to the graphics. Metallized OPET and BOPP films are replacing aluminium foil laminates in many applications.

Typical applications are:

- Snack foods
- Coffee packaging
- Processed meat
- Candy/confectionery
- Bag-in-box liners: for wines and juices

The development of metallized plastic films for use in packaging is a good example of an environmentally friendly material. The aluminium layer deposited is only around 0.05 µm in thickness on a low gauge plastic film, such as 12 µm OPET. Studies have shown that the metallization process uses 0.092 MJ/m^2 of energy compared to 4.2 MJ/m^2 for the production of 7 µm aluminium foil, i.e. approximately 2% of the energy requirement.

Table 11.8

Effect of metallization on barrier properties

FILM (optical density 2.0-2.4)	MVTR (g/m^2.day)		OTR (cm^3/m^2.day)	
25 µm	Clear	Metallized	Clear	Metallized
PA-6	290	7	36	1
OPET	20	1	73	1
BOPP	4	1	2306	51

An aluminium foil laminate as shown below can be replaced by a metallized OPET structure.

OPET (12 µm)
Al foil (7 µm)
LDPE (45 µm)

Metallized-OPET (12 µm)
LDPE (45 µm)

This structure will save ~4 MJ/m^2 of energy by replacing the foil. The calculation is shown in **Table 11.9**.

A 76% savings in total energy is achieved by replacing the 7 µm foil with the metallized film. Similar savings can be shown with 20 µm BOPP. The latter film consumes an estimated 1.9 MJ/m^2 of energy during manufacture. Clearly metallization is a low energy consuming process and also reduces material usage with a sub-micron metallic layer required for adequate barrier.

Table 11.9
Energy consumption of metallized laminates

Energy usage	MJ/m²	
12 μm OPET	2.0	2.0
Metallization	0.092	-
7 μm Al foil	-	4.12
TOTAL	2.092	6.12

Studies have shown that metallized films and laminates can be incinerated in a similar way to equivalent non-metallized films and that the very thin aluminium deposit is easily oxidized to a non-toxic ash. An aluminium foil lined laminate will need agitation in order to break up the metal and will produce higher amounts of ash for a given surface area.

Experience to date with metallized single polymer films, e.g. BOPP/met-BOPP for snacks, has shown that these can be recycled via extrusion in a similar manner to a non-metallized equivalent. In multi-polymer structures such as met-OPET/PE, recycling is more difficult, but this can also be done by an extrusion process, if the recycle film is co-mingled with polyolefins at no more than 5% addition. In general, constraints on the use of recyclates are in the availability of viable applications and markets rather than technical limitations.

11.11 Oxide (ceramic) coated films

Aluminized gas diffusion barriers are now widely accepted by the packaging industry throughout the world. More recently attention has been focused on transparent thin film barrier materials based on oxides, e.g., silicon oxide (SiO_X). These are designed to provide the same performance as their metallized counterparts. These transparent webs are required to have high barrier, be heat and moisture resistant, be **microwaveable** and allow direct measurements with metal detectors on finished packages. The vacuum deposition process is sometimes referred to as ceramic or glass coating.

The deposition under vacuum of silicon oxide or other oxide barrier coating material onto a web, using a commercially viable process is complex. The technology involves an understanding of the subtle interactions between the deposition process, the gaseous environment, the coating thickness, the substrate and the many methods of conversion into a final package. Three vacuum deposition processes can be used to coat silicon oxides of the generic SiO_X type.

1. *Evaporation*, where silicon oxide is placed in a crucible and vaporized with electron beam gun heating

2. *Chemical plasma deposition (CPD)* uses gaseous organosilanes as monomer and the plasma is activated by electromagnetic discharge
3. *Sputtering* using silicon (Si) as the target and oxidizing to SiO_X

The most widely used commercial process is evaporation with electron beam. The high vacuum plants used were developed from the well-established metallization processes and can be used with other oxides (Al_XO_Y). The interest in aluminium oxide is in the lower cost of the aluminium wire. The main drawback with the evaporation process is some yellowing of the coating and relatively poor resistance to flexing.

The chemical plasma technology appears very attractive since it uses gaseous organosilanes as monomers and requires less energy for deposition. The deposits are believed to interact on the film's surface forming thinner and more flexible coatings. However, the process is believed to be difficult to control and operate efficiently. Yellowing is avoided.

The sputtering process suffers from low deposition rate and high power requirements and is rarely used.

Oxide coated films are usually based on OPET or BOPP although other supports, such as polyamide (PA), PE and paper are produced. The films are usually corona or flame treated to improve adhesion. BOPP films are the most difficult to coat.

Practical experience has shown that these silica coated structures can increase shelf life by 10-20 times. There chief advantage is that the films are transparent and microwaveable and easily recyclable. The commercially available films are based on the following supports and will usually be supplied as laminates with other films. The silica coating is typically 200-600 Å thick depending on the deposition process.

- 12 μm OPET
- 20 μm BOPP
- 15 μm OPA

The films can be adhesive laminated or extrusion coated using similar techniques to metallized films. Converting on high speed processes must avoid stretching beyond 2%, which may lead to dislocation of the thin coating with loss of barrier. This limits their use to highly oriented (non-stretch) supports. The high temperatures used in extrusion coating can cause strain relaxation of the oriented film, leading to stretching, which can cause a loss of barrier. The plasma deposition process is claimed to work well in extrusion coating.

Table 11.10 shows some typical barrier properties of 12 μm OPET films. The barrier values are similar to those obtained with metallization. The oxygen and moisture barrier improve by some two orders of magnitude over the basic OPET film.

Table 11.10

Typical barrier properties on OPET films

OPET film 12 μm	Coating thickness Å	OTR (cm^3/m^2.day)	MVTR (g/m^2.day)
Uncoated	0	220	120
E-beam evaporation	600 (60 nm)	3.0	2.0
Plasma deposition	200 (20 nm)	1.0	1.0

There is, however, no commercial advantage over metallized films, but these products are less costly than laminates based on EVOH for equivalent applications. These ceramic coated films have superior barrier and aroma retention properties compared to EVOH based laminates. They will also resist pasteurization and retorting, (heat sterilization) processes. Applications are as laminated structures for packaging dried foods, snacks, soups, toiletries and medical use.

Advances in deposition processes have enabled the application of metallic layers, a few angstroms in thickness, to act as a primer to enhance adhesion in the extrusion coating process. LDPE coated at 310 °C can achieve excellent adhesion on these films. The use of EAA, EVA and ionomers will enable temperature reduction to 280 °C. Coating thicknesses are typically 20 μm.

Some manufacturers are offering ready-to-use films based on oxide-coated 12 μm OPET with an EVA sealing layer 35 μm thick. Typical barrier properties are shown below. These barrier values are excellent considering the very low deposition thickness achieved by this process:

- OTR 0.30 cm^3/m^2.day
- MVTR 0.06 g/m^2.day

It has also been claimed that high density atmospheric plasma can be adapted to deposit silica or other oxides in-line onto moving webs. If validated commercially this may be an additional method of producing high barrier films.

Chapter 12

Multilayered film applications

The flexible packaging industry requires products that must meet a number of criteria, which will be rigorously specified by the application and market demands. No single film will satisfy all requirements. The converter must, therefore, resort to developing multilayered structures that will embody the specified attributes at the lowest cost.

Table 12.1 lists some of the key properties required in high value-added packaging films and the best film type for each.

Table 12.1

Properties required for packaging

Property	Best product
Low cost	LL/LDPE, PP, HDPE, EVA, Paper
Yield (m^2/kg.)	OPET, BOPP, 7 μm foil, metallized films
Stiffness	OPET, BOPP, OPA, paper/board
Flexibility/crack resistance	EVA, LLDPE, mPE*
Durability, puncture resistance	PA, OPET, ionomer
Heat sealability	LD/LL/mPE, EVA, ionomer, EAA
Point of sale appeal	Reverse printed OPET, BOPP, metallized films
Barrier (OTR, MVTR, aroma)	PVDC, EVOH, al foil, metallized/oxide films
Resistance to grease, fats and alcohols	EVOH, PA, HDPE, ionomer
Minimum scalping	OPET, PA, EVOH, all metallic films
Deep draw thermoforming	PA, ionomer, PVC, HIPS
Non-stretch	OPET, BOPP, OPA
Clarity/transparency	OPET, BOPP
Machineability	OPET, BOPP, OPA, paper
Printability	OPET, PA, BOPP, al foil, paper
Food legislation compliance	Must comply according to region and food type

* *metallocene PE*

The barrier requirements for some food products are shown in **Table 12.2**.

Table 12.2

Barrier requirements for some food products

Food product	Property	OTR	MVTR
Snacks, potato chips	Crispiness, aroma and O_2 protection	5-10	0.3-0.5
Sweets, chocolates	Low moisture and aroma loss	10-20	1.0-1.5
Soft cheese	Low moisture and aroma loss	1-3	0.5-1.0
Prepared meals	Low moisture and aroma loss. O_2 protection	0.5-3.5	4.0-5.0
Fresh meat	Low moisture and aroma loss. Colour protection	40-60	1.0-1.5
Frozen food	Minimize moisture loss	1000	1.0-1.5
Soup powders	Low moisture and aroma loss. O_2 protection	1.0	1.0
OTR cm^3/m^2.day @ 75% RH *MVTR g/m^2.day @ 85% RH*			

The following methods are utilized to produce these complex composites. Selecting between the different processes will depend on the properties required at the lowest cost, but also factors such as run length and the availability of equipment and expertise.

Extrusion coating is excellent for paper/board, aluminium foil with low coating weight polyolefins at high line speeds and no drying. Coextrusion options can provide skin layers.

Coextrusion provides great versatility with a wide choice of polymers. Additional tie-layers may be required. Thin barrier layers are possible. Achieving good layer distribution and interlayer adhesion can be challenging.

Adhesive lamination is very versatile for all film to film combinations. Complete curing of reactive adhesives is essential. Some adhesive formulations may have a short pot-life and curing times of several days may be required.

Other coating based on lacquer, solution and latex. Low coating weights and multiple passes may be necessary.

FFS (form, fill and seal) machines are used to package most types of consumer products as liquids, powders or granulates. These include milk, beverages, condiments, cereals, spices, frozen produce and snacks amongst others. The multilayer films reviewed in this section are based on a polyolefin layer laminated or coextruded with other materials, such as OPET, BOPP, aluminium foil, polyamide, metallized films and barrier polymers.

The polyethylene films utilized in flexible packaging are typically used as the heat seal layer as well as adding bulk at minimum cost. For goods requiring moisture protection HDPE with density >0.955 g/cm^3 is used in coextrusions with other polyethylenes and copolymers depending on the required heat seal specifications. The metallocene PEs (mPE) provide outstanding heat seal and low extractables.

The trend is to down-gauge in this highly competitive market to reduce cost and carbon footprint. Blending of polyethylenes is usual to optimize cost, processability and performance.

Film impact strength is best with EVA, ionomer and mLLDPE. It is also acceptable in many applications with LDPE and LLDPE, but poor with HDPE films unless biaxially oriented.

Taint and odour are influenced by the quality of the PE grade, the extrusion conditions (temperature and oxidation) and by the level of corona treatment. LDPE contains no catalyst residues or stabilizers and is usually the most organoleptically neutral of the PE family. Metallocene PEs with their narrow composition distributions (low extractables) are also expected to be exploited in the most demanding applications.

A coefficient of friction (COF) of 0.2 will meet the demands of most packaging machines. The COF is controlled by the precise amount of slip and anti-block additives present and the nature of the surfaces. Copolymers, including Ziegler-Natta LLDPEs, need a higher concentration of slip additives then LDPE homopolymers.

Transparency of the films depends on resin type, extrusion conditions and cooling rate. Ionomers and mPE (metallocene PEs) provide low haze films from the tubular extrusion process.

12.1 Calculating film stiffness

The stiffness of films used in FFS converting processes is an important characteristic for trouble-free operation. The film should ideally be non-extensible and have sufficient rigidity to be pulled and wrapped without wrinkling and creasing. With the current emphasis on down-gauging the risk of producing films that will stretch or be too flexible for machine handling is increased. In coextrusions placing the **higher** modulus materials on the outside will maximize the overall rigidity of the structure.

The stiffness or flexural strength (S) is defined by the product of the elastic modulus (E) and the moment of inertia (J).

$$S = E \times J \tag{12.1}$$

The moment of inertia (J) is calculated from equation 12.2 using the thickness and width.

$$J = \frac{w \times t^3}{12} \tag{12.2}$$

w = width t = thickness

Thus film stiffness (S) is calculated as:

$$S = E\frac{w \times t^3}{12} \tag{12.3}$$

Equation 12.3 shows that the stiffness (S) is proportional to the elastic modulus (E) and the **cube** of the specimen thickness (t). This relationship clearly demonstrates the importance of thickness in determining the overall stiffness, machineability and handle of films. Typical moduli and stiffness values of some PE films at 30 µm thickness and 1 m width are shown in **Table 12.3.**

The stiffness of the plastomer type mPE is close to that of a low EVA (5%VA) copolymer. The zinc ionomer is similar to LDPE and the sodium type is stiffer than LLDPE. HDPE is almost an order of magnitude higher than the resins at the low end of the scale in this table. In **Figure 12.1** the stiffness of a number of heat seal resins is plotted against thickness.

Table 12.3

Stiffness of PE films (30 µm)

Resin	E-modulus (MPa)	Stiffness (cN/mm)
mPE (0.905 g/cm^3)	110	2.5
EVA (5% VA)	130	2.9
Ionomer (Zn^{++})	160	3.4
LDPE (0.919 g/cm^3)	160	3.6
LLDPE (0.918 g/cm^3)	200	4.5
Ionomer (Na^+)	280	6.3
MDPE (0.940 g/cm^3)	550	12.4
HDPE (0.952 g/cm^3)	800	18.0

For a value of 40 cN/mm the Na^+ ionomer will require a film thickness of 55 µm. The mPE (D 0.905 g/cm^3) will in contrast require a film of 75 µm. This variation will affect costs, and is obviously another important consideration when selecting a material for its heat seal properties.

The thinner the film the faster is the sealing and cooling rate. It is also important to bear in mind that the elastic modulus will depend on crystallinity and the method of processing. For instance cast films will be less rigid than blown films. Orientation can also increase the elastic modulus by around a factor of 2.

HDPE is a useful material to increase stiffness by either blending or coextrusion with LL/LDPE. For maximum stiffness HDPEs with densities >0.955 g/cm^3 should be used. Some examples of demanding flexible packaging applications are reviewed below.

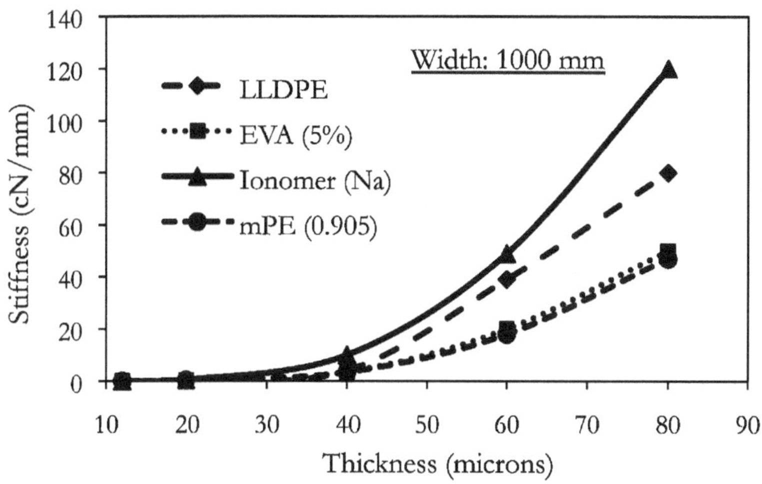

Figure 12.1

Stiffness vs. thickness

12.2 Form, fill and seal packaging (FFS)

One of the most versatile methods of packaging solids, powders and liquids with flexible films is by the use of form, fill and seal (FFS) techniques. These processes are fast, economic and hygienic for edible products. Many of the films used in FFS packaging are based on multi-layered structures produced by coextrusion, lamination and/or extrusion coating.

FFS processes produce flexible packages from roll-fed stock. PE resins play an important role since they contribute the all important function of forming strong heat seals on which the package's integrity will depend. Supermarket shelves are inundated with examples of goods packaged on FFS machines.

There are two basic types of form, fill and seal processes.

1. Vertical form, fill and seal (VFFS)
2. Horizontal form, fill and seal (HFFS)

Figure 12.2 shows the basic layouts of the HFFS machines. There are two set-ups the first is based on a single web, which is folded as it moves downstream and filled. The formed pouch is 3-side sealed. The second method uses two webs, which cover the product and are 4-side sealed. This technique works very well with thermoformed trays, which are produced in-line, and are easily filled and then lidded with the second web.

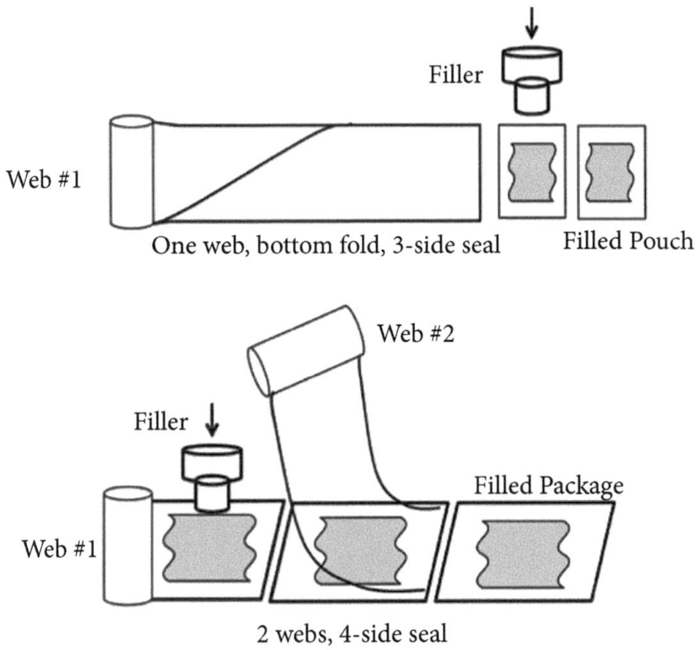

Figure 12.2

HFFS packaging layout

VFFS processes are excellent for the production of simple pillow style pouches or gusseted bags. The latter create more volume and maintain a rectangular shape after filling. The web is folded to form a tube and sealed vertically where the edges meets. The vertical seal can be a fin or lap seal or gusseted as shown in **Figure 12.3**. The type of seal used places different demands on the properties of the film. A gusset will require the heat from the sealing bar to pass through four layers of film, which will slow the process unless a very low seal initiation temperature (SIT) material is used. In the case of the lap seal it is critical that both the inside and outside of the film are sealable to each other. Although not a problem with monofilms this may create difficulties with asymmetric multilayer structures.

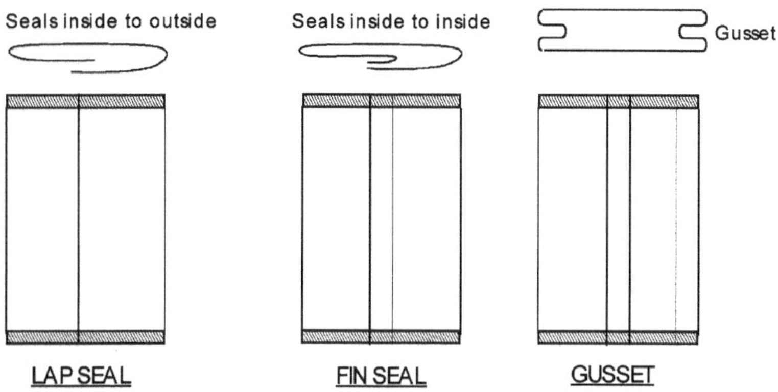

Figure 12.3

Seal types in VFFS systems

Figure 12.4 is a schematic of the basic VFFS layout using a single web. The operational steps are demonstrated as the tube is folded and sealed vertically to form the tube:

1. Filling tube
2. Forming collar to fold web
3. Tracking roller, which feeds the web
4. Vertical sealing bar, which forms the tube
5. Web transport belts
6. Horizontal seal bar
7. Cutting knife

In the final step at 7 the seal bars at the bottom are removed and the pouch or carton is cut and separated. There are variations on this basic process, which use two webs to form the tube section. These packs are 4-side sealed.

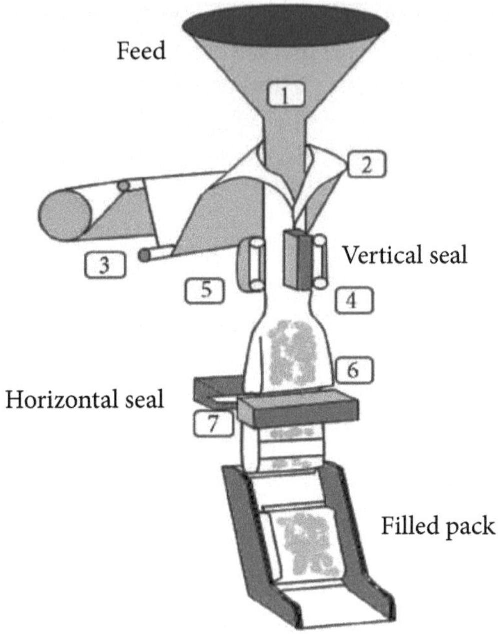

Figure 12.4

VFFS Packaging machine with single web

To avoid opening the bottom of the pouch as it fills under gravity sufficient time must be allowed for the hot seal to cool. It is, therefore, important for the melt to have high enough hot tack to resist opening from the applied stress during filling. Laminates that include an aluminium layer will have a much faster cooling rate enabling higher throughputs.

High frequency (HF) sealing can be used with foil bearing structures particularly with thick paperboard supports. The conductive metallic layer is heated rapidly by induction and melts the sealant layer. This technique by-passes heat dissipation through the thick paper and accelerates the process. Cartons used in liquid packaging will normally use HF sealing.

The VFFS process is low cost and is very versatile in terms of pack sizes and product changes. Production speeds of 300 pouches/min are achieved with some machines. However, the process is more restricted in the choice of width to length ratio of the pouch compared to HFFS packaging.

All the PE polymerization processes based on organo-metallic catalysts will produce polyethylenes with narrow molecular weight distributions. These will convert into films with superior seal strength, and hot tack compared to branched

LDPE and EVA. Lowering the density < 0.915 g/cm^3 with metallocene catalysts will lower seal initiation temperature proportionately.

For the converter it is important to establish whether these new PEs will provide real benefits in terms of faster packaging rates, reduced leakers or reductions in other packaging failures. The key is not to be influenced by promotional claims leading to over-engineered products particularly with complex coextrusions.

In applications where sealing through contaminants such as bacon fat and fine powders is challenging, the ionomers have generally proven to be the material of choice. Ionomers in combination with polyamide result in films with excellent deep draw thermoformability, penetration resistance to sharp objects and grease resistance.

The following critical polyolefin film properties are applicable to form, fill and seal processes for both liquid and dry goods packaging:

- Hot tack strength
- Seal strength
- Low seal initiation temperature (SIT)
- No pinholes
- Low COF between film and metal
- Tear and puncture resistance
- No taint/odour (food packaging)
- No light transmission for dairy products
- Softness and flexibility (pillow packs)
- Peelable seal (easy opening)
- Seal through contamination (fat, powders etc.)
- No static (dry goods, fine powders.)
- Low moisture vapour transmission rate (MVTR) for cereals

For the best results on all FFS machines, a narrow film thickness distribution is required. Recommended thickness specification could vary by ±8% from point to point, and the average value should equal the target thickness.

12.3 Aseptic packaging

Aseptic packaging is defined as the filling of a sterile product into a sterile container and hermetically sealing the container so that infection is avoided. The packed perishable product can be shelf stable for over six months. A wide range of edible products such as milk and dairy products, puddings, desserts, fruit and vegetables, juices, soups and sauces are packed aseptically. Where aluminium-lined paperboard packages are used the tandem extrusion coating process is recommended.

The advantage of aseptic packaging is the reduced loss of nutritional and organoleptic properties compared to canning where longer heat treatment conditions are used. Based on paper and plastic films these packages are lighter than tinplate. The high temperature short time (HTST) sterilization process also referred to as UHT (ultra high temperature) is efficient and minimizes loss in taste and flavour. The process flow chart is shown in **Figure 12.5**. The flat web is passed through a hydrogen peroxide bath for six seconds at 70 °C. The residual peroxide is driven off with hot air and the sterile web flows into the filling machine.

The technique is also used to package non-sterile products to avoid invasive bacteria. Examples are fermented dairy products such as yogurt.

Sterile medical and pharmaceutical products are also packed using aseptic processes. Other sterilization techniques, e.g., steam sterilization, electron beam, ethylene oxide and γ-radiation can be used. Steam sterilization is the least energy efficient and is usually avoided.

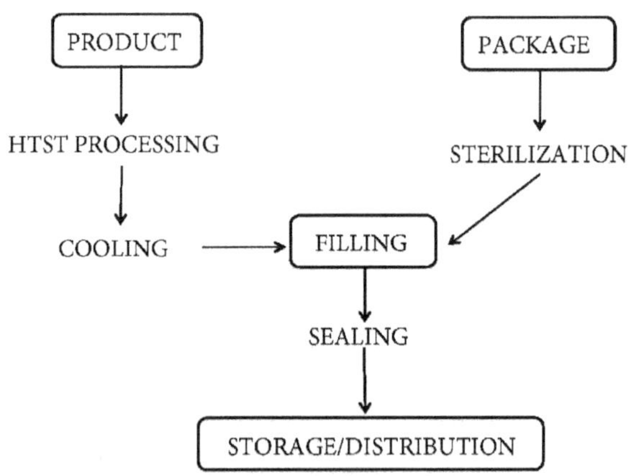

Figure 12.5

Aseptic packaging

12.4 Bag-in-box (BIB) laminates

A bag-in-box (BIB) is a flexible laminate bag, with a dispensing spout or valve, encased in a cardboard box or wooden crate. The system is used to dispense liquids and pourable foods and sauces. Typical applications are in the packaging and dispensing of wines, fruit juice beverages, juice concentrates, soups, tomato pastes etc. BIB packaging is also used in industrial applications such as battery acid, liquid detergents, motor oils and photo-chemicals. Practically every commercial pourable product can be considered for BIB packaging.

Aseptic filling where the food and bag are sterilized separately is widely used. The bag is usually a bag within a bag.

Bag-in-box systems can range from small 2 litre sizes up to 2000 litre bulk containers. The end-user requires seal integrity combined with flexibility and puncture resistance during handling and transportation. In addition excellent flex crack resistance is required and for food products the package must protect its contents from odour, taint and aroma loss. It is important that the bag is sufficiently flexible and able to collapse as fluid is drawn out through the dispenser. The advantages of BIB packaging are:

- Barrier protection from light and oxygen
- Longer shelf life after opening: up to 8-10 weeks
- Low resource usage (10 g/litre) and recyclability
- User-friendly and tamper-evident
- Aseptic/non aseptic, hot/cold filling possible
- No cleaning costs
- High brand recognition
- Prevents oxidation in wine packaging

Critical properties are:

- Sealing (low initial seal temperature, through contamination)
- Hot tack strength at low heat seal temperatures
- Puncture resistance
- Compliance with food packaging legislation (for non-industrial uses)
- ESCR (environmental stress crack resistance)
- Softness and flexibility: bag must collapse on emptying

- Low taint/odour/scalping
- Pinhole free

The inner bag includes the tap or dispenser and is sterilized by gamma (γ) irradiation. The moulded dispenser is welded to the bag and must be absolutely leakproof. The inner bag is constructed from 50-100 μm LL/LDPE. The second (outer) bag provides cushioning and additional protection from leakers. When gas barrier is specified the outer bag is usually based on a metallized OPET laminate with a seal layer. Seal layers are usually based on blends of LL/LDPE and EVA.

Some applications may require more than the 2-ply construction for the outer bag. Traditionally the inner ply consists of a tough, high strength seal resin, such as EVA copolymer of 9-12% VA content. However, LLDPE very often blended with EVA is rapidly penetrating this market because of its better ESCR, toughness, seal and impact strength. The increased cost of higher α-olefin LLDPE or metallocenes can be compensated by down-gauging the film.

Heat sealable high gas barrier films for BIB applications are produced by:

- Lamination of **metallized** polyester (OPET) with EVA sealant
- Lamination of PVDC-coated OPET with EVA sealant
- PA film and metallized PA film, with EVA as sealant
- Coextrusion: LLDPE/EVOH/LLDPE

For wine packaging, metallized OPET adhesive laminated with EVA (5-9% VA) or LLDPE is the most common structure. Metallocene PEs with their advantage of very low extractables and outstanding seal strength are particularly appropriate for this market sector.

Special grades of EVOH with high flex crack resistance are available. These will usually be specified with high ethylene content. The moisture sensitivity of EVOH may limit the wider use of these structures. EVOH can be coextruded with polyamide 6 without a tie-layer. This combination has outstanding barrier and very low aroma loss. Laminating this structure to polyethylene can provide an outstanding film of high strength and flex crack resistance. Metallized films and aluminium foil laminates may crack on flexing and may have poor resistance to acid foods (pH < 6.4).

A typical film laminate for the outer bag would have the following structure:

EVA (50 μm)
metallized OPET (12 μm)
EVA (50 μm)

Large bags (1000-2000 L) for special applications such as beer, wine storage and dispensing are likely to be based on a structure such as the following:

Inner bag	LD/LLDPE bag in contact with liquid
Outer bag	Coextrusion: LLDPE/tie-resin/EVOH/tie-resin/EVA

The tie-resin will be a maleic anhydride grafted PE (MAH-g-PE) for adhesion to EVOH.

12.5 Dry food (cereal) packaging

Breakfast cereals are one of the most demanding types of dry foods to package. These products are processed and packaged to achieve a shelf life of one year after production. HDPE is the best moisture barrier. Highly flavoured cereals require additional gas and aroma barriers for protection. Polyamide (nylon) is considered the best aroma barrier.

Film thickness ranges from 38-75 μm. The EVA seal layer accounts for 17-20% of the total film weight. Typical barrier requirements for cereals are:

- MVTR: 3.5 g/m^2.day @38 °C, 90% RH
- OTR: 100 cm^3/m^2.day @ 23 °C, 0% RH

Some applications may require a MVTR below 2 g/m^2.day, and a typical structure is HDPE (40 μm)/EVA (8 μm). Ionomer is sometimes preferred as the heat seal layer, and for seal peelability a 50/50 blend of sodium and zinc ionomer can be used.

Price competition is continually forcing converters to down-gauge by evaluating the most effective moisture barrier materials. Two coextruded structures used as box liners for cereal and other dried foods are shown in **Figure 12.6**. The thickness and type of HDPE will be critical in determining the moisture resistance of the pack.

The coextruded breakfast cereal box liner film is typically 35-45 μm in thickness with volume ranging from 2-3 litres. The second structure in **Figure 12.6** combines HDPE with polyamide (PA6) for additional moisture and aroma barrier. The tie-resin is a maleic anhydride grafted PE. The polyamide films as stated elsewhere embody outstanding toughness and oil resistance.

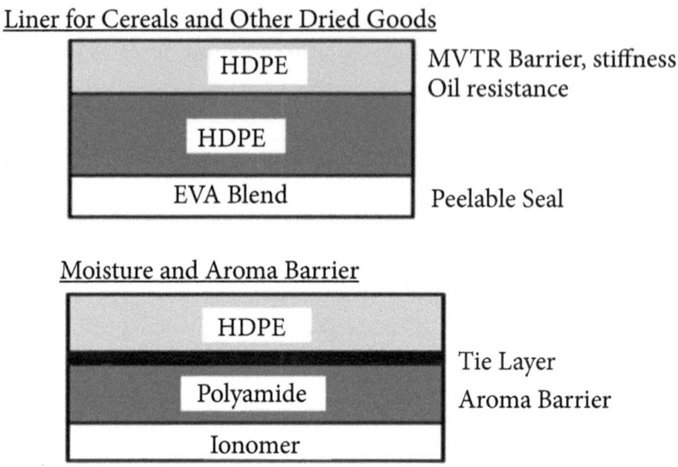

Figure 12.6

Coextruded films for FFS applications

12.6 Snack foods

Snack foods are typically roasted, fried, salted and flavoured products often consumed between meals. Snack foods are highly flavoured and contain fats requiring greaseproof packaging. To retain the texture (crispness) the moisture content must be kept very low. Low MVTR of the package is essential.

Shelf stability specifications can range from 2 to 16 weeks.

Typical examples are:

- Potato crisps
- Peanuts
- Cashew nuts
- Dried fruit
- Almonds
- Cheese balls
- Nutrition bars
- Wafers

This is a high value-added market where the products can be affected by moisture, oxygen and flavour changes. Fried snacks in particular are spoiled by the presence of oil, which will lead to oxidative rancidity. These reactions are accelerated by heat, moisture and air.

Snack foods are usually packed on automatic VFFS machines at high filling rates. Where fat content is high flushing with nitrogen is required.

Typical flexible film structures are listed below. BOPP and OPET films are widely used. Metallized 12 μm OPET is the most widely used barrier film. These laminates are usually produced by adhesive lamination. The adhesive must be resistant to attack by oils and fats. The polyolefin heat seal layer will range from 35 -80 μm in thickness. Some applications specify 155 μm seal layers. The seal layer can be blends or coextrusion based on LD/LL/HDPE. EVAs and ionomers are also used.

- BOPP/LDPE
- BOPP/OPET/ LDPE
- Metallized OPET/LDPE
- BOPP/metallized OPET/LDPE
- OPET/LDPE
- OPET/Al foil 9-12 μm/LDPE

Other structures are based on a PVDC barrier, where the PVDC provides heat sealability.

- PVDC coated OPET
- BOPP/PVDC/BOPP
- PVDC coated glassine
- PVDC coated glassine/BOPP

The films must be easily printable as brand identification is crucial in this market.

12.7 Films for coffee packaging

Coffee packaging requires very high performance laminates and the following properties are all important:

- Puncture resistance
- Flex crack resistance
- Dead-fold

- Seal through contamination
- No extractables
- No flavor loss
- No static

In vacuum packs outstanding gas barrier is also essential. Low static films should be used with coffee and other fine powders. Ethoxylated amines, e.g., nonylphenol ethoxylate, are the most widely used antistatic additives in polyolefins. To be effective these amines must react with atmospheric moisture. Measured surface resistivity or decay time shows that effective antistatic response will occur only in the presence of at least 40% RH. A resistivity of 10^9 ohm is obtained with ethoxylated amines at the right conditions.

Two lamination examples for soft packs are:

- OPET/PVDC/LLDPE
- Metallized OPET/LLDPE

Metallized films are increasingly being used to replace the aluminium foil layer. The driving forces here are environmental and cost. The replacement of OPET/PVDC by OPET/SiO_x films has been suggested as a means of eliminating the use of a chlorinated polymer. There are still some paper/PVDC coated materials in use, but these are being phased out.

The most common film structure for vacuum packs is based on the generic structure in **Figure 12.7**.

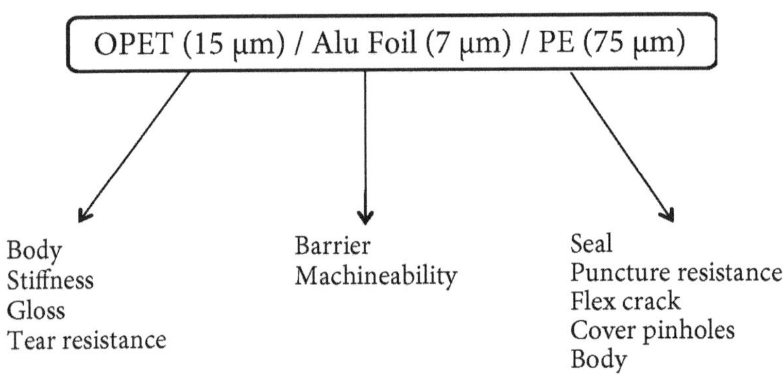

Figure 12.7

Typical coffee packaging laminate

Another innovation is the replacement of the traditional barrier layer with EVOH. A 5-layer coextrusion of LLDPE/Tie/EVOH/Tie/LLDPE laminated to OPET is one solution. EVOH based laminates for vacuum coffee pouches are successful in Japan and have been adopted in Scandinavia and Switzerland for environmental reasons. One of the latest concepts for coffee packaging is shown in **Figure 12.8** where EVOH barrier and metallization are combined and laminated to reverse printed BOPP. The sealability and puncture resistance of the 3 layer PE film can be improved by blending with 20-30% mLLDPE. It is not clear whether the extra cost of combining metallized PET with coextruded EVOH is likely to be cost effective.

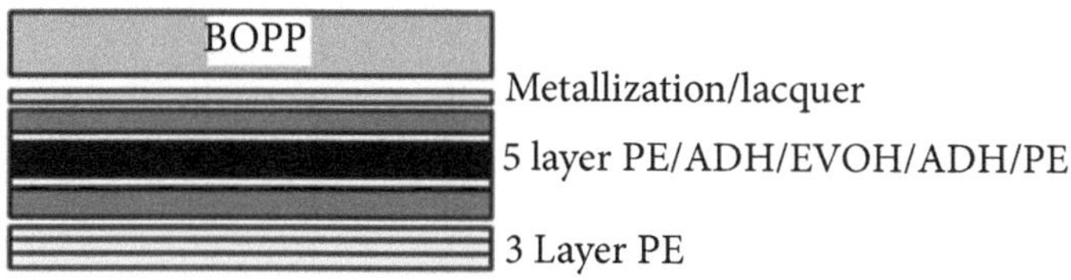

Figure 12.8

Newer vacuum coffee pack

The two principle markets that utilize flexible films for coffee packaging are institutional packs, e.g., coffee machines in offices, and vacuum packed coffee for home use.

The support films are based on BOPP, OPET or OPA. The vacuum packs employ 3-4 ply laminates and the institutional packs are usually based on 2-ply films. Ground coffee is highly abrasive and can readily puncture the pack. For the best puncture resistance oriented polyamide (OPA) films are recommended. In some applications the extremely tough and costly bi-oriented PA (BOPA) is preferred.

Institutional coffee packs. These are usually 2-ply films based on 12 µm metallized OPET laminated or coated with 50 µm LDPE. The OPET film is printed and over-lacquered if necessary for abrasion resistance.

For larger sized packs (2-3 kg) the higher strength metallized BOPA is preferred over OPET. The structure is as follows:

15 µm metallized OPA	
	PVDC coating
60 µm LDPE or LLDPE	

This film offers greater puncture and flex crack resistance then OPET, but carries a premium price.

Vacuum packs. Some typical structures are as follows.

1. 12 μm OPET/8-9 μm Al foil/15 μm OPA/75 μm LD or LLDPE
2. 12 μm OPET/20 μm BOPP/8-9 μm Al foil/85 μm LDPE
3. 12 μm OPET/8-9 μm Al foil/85 μm LDPE

Film #1 is the most popular as it provides for high quality printing on the OPET and outstanding puncture and stress crack resistance from the oriented polyamide film. LLDPE provides superior seal strength and hot tack. Film #2 provides a 12 month shelf life for ground coffee. Structure #3 with its 3 layers has a shelf life of 6-8 months and is apparently the least popular film.

12.8 Frozen food films (retail)

Freezing entails reducing temperatures below zero so that microbiological, enzymatic and bacteriological reactions are retarded. In freezing fresh produce the temperature must pass rapidly through a transition from water to ice so that ice crystals are relatively small and do not disrupt the plant tissue cells. The produce may be frozen either inside or outside of the package. Modern freezing plants use high velocity cold air or liquid nitrogen (N_2) to remove heat from the bulk unpacked or individually quick frozen (IQF) produce. Cooking vegetables are very often frozen in IQF form. The frozen produce is then packed in coated paperboard cartons or PE pouches. Care must be taken to avoid sublimation of ice, which can lead to 'freezer burn'.

Frozen food films have to meet the following requirements:

- Strength/toughness at -20 °C or below
- Outstanding heat seal strength
- Hot tack for produce packed in FFS machines
- Low COF of film to metal (applicable to FFS converting)
- Stiffness to avoid snagging on mandrels etc. (applicable to FFS converting)
- Clarity if non-pigmented
- Printability (high quality)
- Food regulation compliance

Vegetable such as green peas, sweetcorn, broccoli, carrots, spinach, cauliflower etc. are packed in PE bags. The bags are produced from 50-75 μm film. EVA or LLDPE blends are preferred because of their higher impact strength at low temperatures. The EVA will incorporate 2-5% VA and have a fractional melt index. The higher α-olefin LLDPEs will also provide a good combination of properties. Coextruded films have been penetrating this market.

The films are either clear or white opaque (TiO_2) and flexographic printed if required. Owing to possible damage by UV freezer lights and the presence of unsightly ice crystals high levels of TiO_2 pigment are usually blended into the PE via a masterbatch. Packaging is carried out on VFFS machines. Some structures are shown in **Figure 12.9**.

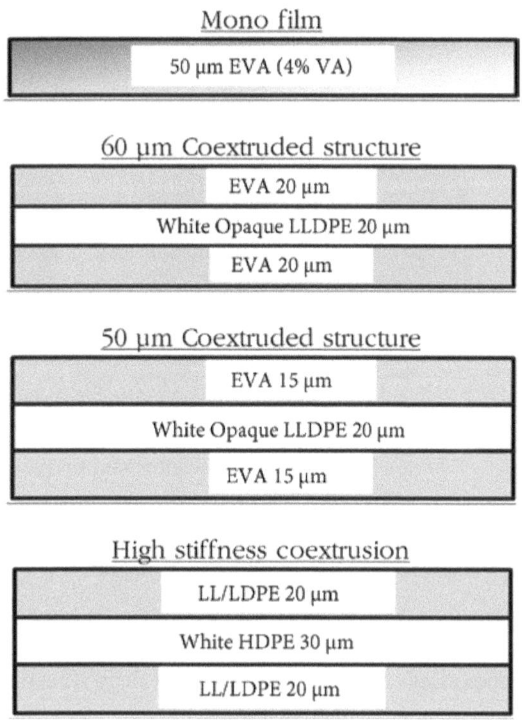

Figure 12.9

Frozen food films

In all of the above structures the metallocene catalysed PEs will play a useful role. EVAs can be replaced by the plastomers-type materials (<0.915 g/cm^3 density) and higher density mLLDPEs can be used to add stiffness. In most cases the metallocenes can be used in blends with either LDPE or EVA. The mPEs will add improved hot tack, higher tensile and superior impact strength. The rationale is to reduce cost and improve processability.

For frozen food films attention must be paid to the strength of the film particularly its seal strength at low temperature. **Figure 12.10** compares the impact strength of films at ambient temperature and at -20 °C. The linear resins have a melt index of 1 dg/min. The LLDPE density is 0.920 g/cm³ and the mLLDPE is 0.916 g/cm³. The relatively low density of the mLLDPE used in these measurements partly explains its superior impact strength. The data shows that the octene LLDPE and EVA are insensitive to the two test temperatures.

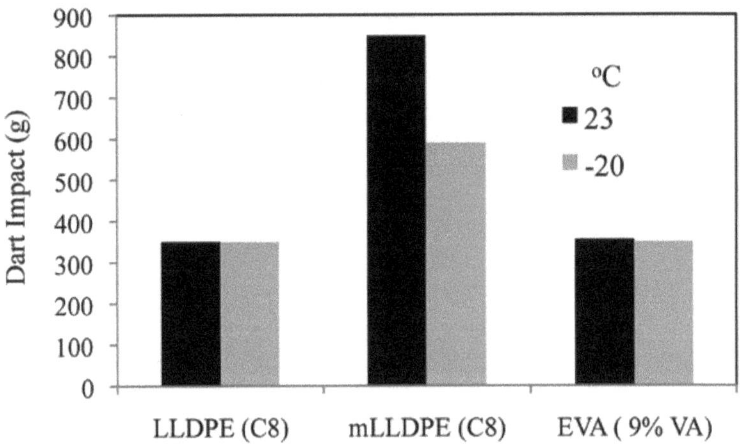

Figure 12.10

Film impact strength at 2 temperatures

Physical property measurements on a 3-layer coextruded structure are shown in **Table 12.4**. The drop tests were carried out at -20 °C with filled pouches (500 g).

Table 12.4

Properties of coextruded films for deep freezing

3-layer coextruded film		
LL/LDPE blend	HDPE (white)	LL/LDPE blend
15 μm	30 μm	15 μm
LDPE: MI 1.2/D 0.922. LLDPE: MI 1.1/D 0.921		
PROPERTY	**UNITS**	**VALUE**
Tensile yield MD/TD	MPa	21.6/16.6
Tensile @ break MD/TD	MPa	33.4/28.0
Elongation @ break MD/TD	%	478/862
Seal strength	N/15 mm	14.5 @ 145 °C
Drop test from 1 m @ -20 °C	# of failures from 10	6
Drop test from 2 m @ -20 °C	# of failures from 10	9

The HDPE layer (MI 0.15/D 0.947) contains 1% TiO_2 added from a 50/50 LDPE masterbatch. LD/LLDPE blends were set at 1 : 1 ratio. The use of HDPE does weaken the low temperature drop tests.

12.9 Liquid pouches

The best example of a demanding liquid packaging application successfully developed with polyethylene films is the milk pouch, first commercialized in Canada and Switzerland and now widely used in India. In Canada 60% of fresh milk is sold in pouches.

The initial concept was based on the use of 2-layer coextruded blown film, which produced tough flexible pouches and crucially the two layers eliminated the risk of pinhole formation. The later availability of LLDPE further enhanced properties and in particular heat seal strength, which reduced the incidence of leakers. This application was one of the first successful uses for LLDPE. LLDPE, particularly octene grades, dramatically reduced leaker rates of the pouch compared to LDPE. Their higher tensile and impact strengths also enabled cost cutting by down-gauging.

The polyethylene pouch is one of the most cost effective and low carbon footprint methods for packaging liquids. A typical pouch weighs 7 g and will hold 1 litre of fluid, i.e., ~0.7% by weight of packed product. The original milk pouch structure was based on 2-layer coextruded blown film embodying a 50-60 μm white and 20-30 μm black or brown inner layer for UV protection. Over the years variations on the basic design have been developed, such as fitting pouring spouts and various stand-up designs.

India is one of the largest users of PE pouches for milk packaging. Initially these were based on 75 μm LDPE film. By the 1980s these were replaced by 65 μm LLDPE blended with LDPE. Down-gauging has continued to 55 μm using improved LLDPE blends. Packaging is typically carried out at a rate of 50/min. The chilled pasteurized milk is consumed within the day of purchase.

Typical specifications are based on 40% LLDPE blended with LDPE:

- Density 0.918 - 0.922 g/cm^3
- Melt index 0.8 - 1.5 dg/min
- Slip (COF) < 0.3

Strong leakproof seals are essential to minimize wastage.

In many markets, particularly Europe and the USA, consumers prefer the convenience of plastic bottles or paper cartons over flexible pouches. Unlike bottles the pouches are not re-closable. However, there has been a renewal of interest for pouches as a way to reduce packaging waste. The key was the development of user friendly systems that enable the pouring of the milk without spillage. Systems are now available with a specially designed jug, which holds and punctures the pouch for pouring.

A similar idea was the development of the stand up pouch to mimic the metal can or bottle. A laminate typically OPET/PE, is sealed along three edges and the narrower side is gusseted usually with a round mandrel. The gusset allows the pouch to be expanded to form an elliptical bottom, enabling the pouch to stand up. The open end forms the filling port of the pouch followed by closure and heat sealing. Below the seal a zip closure can be fitted. Pouring spouts can also be welded on for easy delivery of the packed fluid. **Figure 12.11** shows an open plan view of the laminate. The gusset is formed at one seal, which opens up to form the bottom of the pouch. Seals of 7 mm width are shown in this sketch.

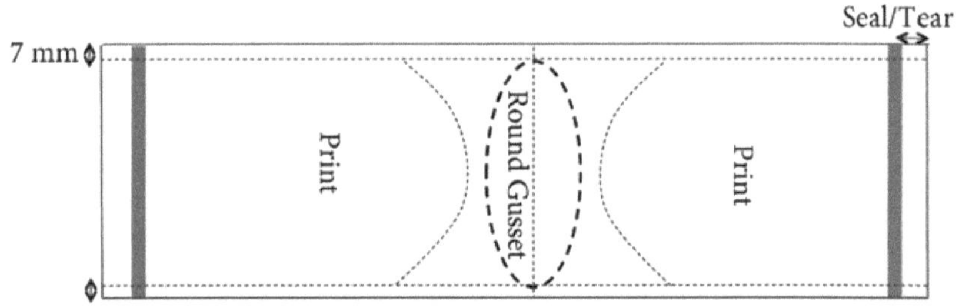

Figure 12.11

Plan view of stand-up pouch

Figure 12.12 shows a typical stand up pouch with a re-sealable zip closure. Sizes can vary from 250 ml to 4 litres. A wide range of laminates are used, which can include paper, plastic films and metallized structures.

Figure 12.12

Stand-up pouch

There has been some interest in the US in replacing PET water bottles with pouches. The aim is resource reduction and lower carbon footprint. These are stand up pouches made from laminates of OPET film with polyethylene. On emptying the pouches collapse and occupy much less space. Because of its blandness, water requires an organoleptically neutral package to avoid imparting any off-taste. Selection of the seal layer in contact with the water is critical.

Another innovation in the US is the packaging of wine in pouches. These designs can hold the equivalent of the standard 750 ml glass bottle. Larger sizes are possible. The pouch is re-sealable and can include a pouring spout. These will presumably compete with the well proven bag-in-box systems.

12.10 Modified atmosphere packaging (MAP)

The success of modified atmosphere packaging (MAP) is driven by the ability to maintain freshness and quality of food produce over an extended time period without the use of preservatives and stabilizers. Increased shelf life reduces waste and facilitates stock control in retail outlets.

MAP is based on flushing the package with a controlled mixture of nitrogen, oxygen and carbon dioxide. A vacuum may be applied to ensure that the pack is effectively flushed. The finely balanced gas mixture slows down the aging process to reduce spoilage, rancidity, mould, flavour changes and other anaerobic reactions.

For living plant tissue the technology is based on the reverse photosynthesis that takes place once a living plant is harvested and pre-cut. The respiratory process is reversed with the plant taking up oxygen and exhaling carbon dioxide. MAP must equilibrate the respiratory gases inside the package. The film must allow restricted passage of oxygen (O_2) **in** and carbon dioxide (CO_2) **out**, until equilibrium is reached between the respiration of the produce and the diffusion of the gases. The rate at which oxygen is taken up while carbon dioxide, water and compounds (e.g. ethanol and aldehydes) which cause off-flavours are released is defined as the respiration rate. The plant will also lose water inside the sealed pack forming condensation. The packaging film must, therefore, allow sufficient moisture to escape to minimize mould formation, but the loss must not be excessive and allow the produce to dehydrate. The respiratory process is illustrated in **Figure 12.13**.

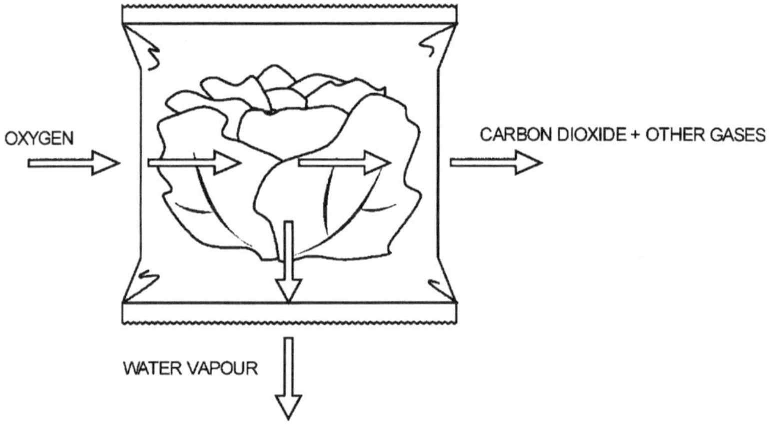

Figure 12.13

Respiratory process

The MAP market is divided into two main sectors:

Food service typically requires a shorter shelf life and uses large packs (3-12 kg). Aesthetic requirements such as print quality, clarity, gloss are minimal. MAP has caught on rapidly in the food service industry with the realization of cost savings obtained with centralized processing, reduced spoilage and improved product quality.

Retail outlets. Smaller packs (0.3 to 2 kg) are used in this sector. Attractive aesthetics are required particularly sparkle, low haze and stiffness associated with consumer perception of freshness. Turnover is slower than in the service sector requiring extended shelf life. Shelf-lives of up to 21 days have been reported for certain vegetables.

Different plant tissues have different respiratory rates, which must be taken into account when designing the packaging system. Preparation such as slicing and rasping will increase surface area, which accelerates the respiratory rates significantly and must be allowed for in the package design. The ratio of the film surface area to produce weight will also influence the balance of respiratory gases inside the pack. A large 5 kg pack will require a different permeability film to a smaller ½ kg pack for the same produce. For some plants if the film permeability is too high then oxidative degradation will take place, whereas if it is too low anaerobic browning will occur. Therefore, the film's gas transmission rate (GTR) has to be optimized for each application.

The respiratory rates are determined by measuring the rate of carbon dioxide (CO_2) exhalation per kilogram (kg) of produce. Some values are reported in **Table 12.5** at two temperatures.

Table 12.5
Typical respiratory rates

Produce	mg CO_2/kg @ 10 °C	mg CO_2/kg @ 15 °C
Tomatoes	12-18	16-28
Bananas (green)	5	21-23
Cucumbers	23-29	24-33
Iceberg lettuce	21-40	32-45
Cauliflower	32-36	43-49
Salad	32-46	51-74
Radishes	31-36	70-78
Brussels sprouts	63-84	64-113
Broccoli	75-87	161-186
Mushrooms	100	180

The key film properties to be considered in MAP are:

- Gas transmission rate (GTR) of O_2 and CO_2
- Moisture transmission rate (MVTR)

Physical and optical properties:

- Low haze
- High gloss
- Low seal initiation temp. (SIT)
- High hot tack
- Stiffness for FFS processes
- Strength at low temperature

In designing the total MAP system for a given plant tissue the following parameters have to be considered:

- Product weight (W)
- Respiration rate of produce at storage temperature and gas partial pressure (R_{tp})
- Film thickness (t)
- Package surface area (A)
- Ambient gas partial pressure (P_a)
- Optimum internal gas partial pressure (P_{oi})

The film transmission rate (TR_f) is calculated from equation 12.4 (K is a unit conversion constant).

$$TR_f = \frac{K \times \left(W \times R_{tp} \times t\right)}{A \times \left(P_a - P_{oi}\right)} \qquad (12.4)$$

As a guideline it is important to remember that film permeability will change by 4-5% per °C change in ambient temperature.

Unfortunately respiratory data at actual storage conditions is variable and subject to produce variety, harvesting methods, processing conditions and timing. Equation 12.4 is, however, still useful as it allows calculation of the effects of a change of any of the variables on the internal partial pressure (P_a). By measuring the equilibrium headspace gas concentration inside the package with a well defined plant packed in a film with a known TR_f the respiration rate (R_{tp}) of the plant can be calculated. Using this number in equation 12.4 with the optimal gas partial pressure will yield the film transmission rate (TR_f) for any given change in pack dimension or filling weight.

A few types of polymers used in MAP applications are shown in **Table 12.6**. The gas transmission rate (GTR) data were determined at 23 °C. Storage conditions for fresh produce will usually be at 3-5 °C, which will slow the respiratory rate and reduce transmission. Corrections, therefore, have to be made to allow for actual in use temperatures. For polyolefins the ratio of OTR at 5° and 23 °C is ~3. For CO_2 the TR ratio is ~2.

Table 12.6

Comparative gas transmission rates (23 °C)

Polymer type	Density (g/cm^3)	25 μm OTR (cm^3/m^2.day)	25 μm MVTR (g/m^2.day)	25 μm CO_2 TR (cm^3/m^2.day)
PP (Homopol.)	-	3500	5	14000
LLDPE	0.920	7000	10	31000
EVA (9%VA)	-	11000	70	80000
mPE (hexene)	0.900	12750	15	62460
mPE (hexene)	0.910	7500	19	33540
Terpolymer m-PE	0.900	14200	25	75300
PVC (plasticized)	-	11870	473	-

GTR is increased as PE density is reduced, i.e. comonomer increased. All the PE resins shown have higher GTR than PP. EVA tends to form sticky films, which can be difficult to convert. The metallocene PE (plastomer) grades have low extractables, are less tacky and have very low haze. However, the films are highly flexible and lack the rigidity of PP. The consumer preference for rigid sparkling films and the greater difficulty of FFS machines to handle extensible films at high speed may limit the use of plastomer type resins as monofilms. Coextrusion of plastomer with a rigid material is, therefore, an obvious solution to tailor the properties for this demanding application.

The retail sector has traditionally used cast PP/PPRC (random copolymer) coextruded films for MAP of fresh produce. PP films are low cost, have low haze, high gloss, high stiffness and adequate strength and seal performance for most FFS processes. However, the relatively low gas transmission rates (GTR) of PP films has limited shelf life to 3-5 days with most produce. The low OTR of PP films will lead gradually to anaerobic conditions and spoilage caused by oxygen extinction as the respiratory process converts the oxygen into exhaled carbon dioxide. High GTR films were, therefore, evaluated over the years to try and further extend shelf stability by better control of the respiratory process. With coextrusion it is now possible to control the GTR of the film by the choice and thickness of polymers in the structure. From a theoretical basis as shown in equation 12.4, combined with field experience, converters have been able to devise empirical relationships which correlate a plant's respiratory rate with the design and size of the package in order to maximize its shelf life. Other properties such as toughness, hot tack, SIT and clarity can also be optimized by the constituents of the coextruded film.

Some permeability data for LLDPE and mLLDPE 3-layer coextrusion is shown in **Table 12.7**. The coextruded structure is configured as follows:

LLDPE (octene)	15%
mLLDPE (D: 0.902 g/cm^3)	70%
LLDPE (octene)	15%

The metallocene plastomer contains 12.5% octene comonomer (density: 0.902 g/cm^3) and has a melt index of 3.0 dg/min. The film was extruded by a cast coextrusion process, which partly explains the higher than expected OTR value.

Table 12.7

Permeability of plastomers and PVC films

Gas type	Coex. mLLDPE	PVC (plast.)
OTR (cm^3/m^2.day)	22,800	11,870
MVTR (g/m^2.day)	46	473
Elastic recovery (%)	88	89.6

Compared to plasticized PVC the polyolefin film has double the OTR and one tenth the MVTR.

The technology has been extended beyond plant organisms to include fresh and processed meat, fish, cheese, pizza and fresh pasta. Some applications require high barrier films or lidded trays. These are usually based on coextrusions of PP with EVOH.

Chapter 13

Polyethylene materials and processes

Five commercial processes are used to manufacture polyethylenes.

1. High pressure autoclave process
2. High pressure tubular process
3. Gas phase low pressure process
4. Solution low pressure process
5. Slurry (particle form) low pressure process

The bulk of production is split between the three key polyethylene categories is as follows:

- LDPE (low density polyethylene)
- HDPE (high density polyethylene)
- LLDPE (linear low density polyethylene)

Low density polyethylene (LDPE) is defined by a density range of 0.915-0.934 g/cm^3. The polymer has a high degree of short and long chain branching, which inhibits the formation of intermolecular attractions. This limits its tensile strength and modulus, but enhances impact resistance. LDPE is polymerized at high temperature and pressure with peroxide initiators that are incorporated into the growing molecular chain and do not leave any residues in the final product. The polymer has excellent flow properties at moderate to high shear conditions and is easy to process by all extrusion techniques.

Linear low density polyethylene (LLDPE) is defined by the density range of 0.915-0.925 g/cm^3. The polymer contains only short chain branches formed by copolymerization with α-olefins. The polymer has generally superior physical properties to LDPE, but is less sensitive to shear in the melt phase making it more difficult to process. In commercial film extrusion processes sharkskin (surface roughness) can occur at the die lips and its low elongational viscosity will cause instability as the melt exits the die in tubular film and extrusion coating processes.

LLDPE is produced by low pressure processes using multi-site titanium based Ziegler-Natta catalysts and more recently metallocene single-site catalysts. Metallocene catalysts produce narrow composition distribution polymers with superior physical properties and low levels of extractables. The metallocene technology has also extended the density range below 0.915 g/cm^3.

High density polyethylene (HDPE) is defined by a density greater than or equal to 0.945 g/cm^3. The polymer contains a low degree of branching (high crystallinity) and consequently much stronger intermolecular attractions. The polymer has greater rigidity and a higher melting point (130-133 °C) than the lower density materials.

HDPE can be produced with supported chromium oxide catalysts, Ziegler-Natta catalysts or metallocenes. The very low branching level is ensured by using the appropriate catalyst formulation, reaction conditions and control of comonomer (0.5-3%) usually n-hexene. Cascade dual reactor technologies have been developed to produce very wide molecular weight distribution resins at high molecular weight that enable the production of polymers for demanding applications such as pressure pipes, industrial containers and high strength low gauge films.

Medium density polyethylene (MDPE) is defined by a density range of 0.930-0.945 g/cm^3. These can be produced by all catalytic processes at low pressure. MDPE has better stress cracking resistance than HDPE, but a lower tensile strength and stiffness.

Very low density polyethylene (VLDPE) is defined by the density range of 0.880-0.915 g/cm^3. The polymer contains high levels of short chain branching (α-olefins). When produced by Ziegler-Natta catalysts, these products have wide composition distributions, which creates large amounts of unwanted waxy material and generally disappointing properties. These polymers are preferably polymerized with single site metallocenes catalysts. These have narrower composition distribution, which results in films with much reduced extractables and outstanding physical properties and heat sealability.

Ethylene copolymers. Comonomers containing heteroatoms will poison all organo-metallic catalysts. Consequently these copolymers are only produced by the free radical activated high pressure process and incorporate comonomers such as vinyl acetate, acrylic acid and alkyl acrylates. The presence of an acid or ester functionality enhances specific adhesive properties and miscibility with many organic and inorganic substances. The comonomer reduces crystallinity, which reduces stiffness and melting point. These copolymers are easily processed by all film forming techniques.

13.1 Low density polyethylene

LDPE is still today the mainstay of the film and extrusion coating industry. LDPE is easy to process, provides adequate moisture barrier, has good sealing properties and is attractively priced.

LDPE has a density range from 0.915 to 0.932 g/cm^3 and is made at very high pressure and temperature. The polymerization is by random free radical reactions initiated with oxygen or peroxides. Typical peroxides used are benzoyl peroxide, di-t-butyl peroxide and t-butyl perbenzoate amongst others. Once initiated, the polymerization is exothermic. Reactor pressure and temperature will range from 1500 to 2500 bar and 200 to 300 °C.

The higher the reactor pressure the higher the density. Melt index is controlled with temperature, pressure and the use of chain transfer agents. Iso-butene and propylene are widely used chain transfer agents that control the degree of polymerization. The reaction vessels are either stirred autoclaves or long tubular reactors. The autoclave reactors can vary in LD ratio from 5 to 18 : 1. The higher LD ratio vessels can include multiple initiator injection points along their length. This makes it possible to run a temperature profile in the reactor thus allowing more control over molecular weight distribution and the level and type of branching. **Figure 13.1** shows a diagram of a long LD ratio high pressure autoclave reactor for LDPE.

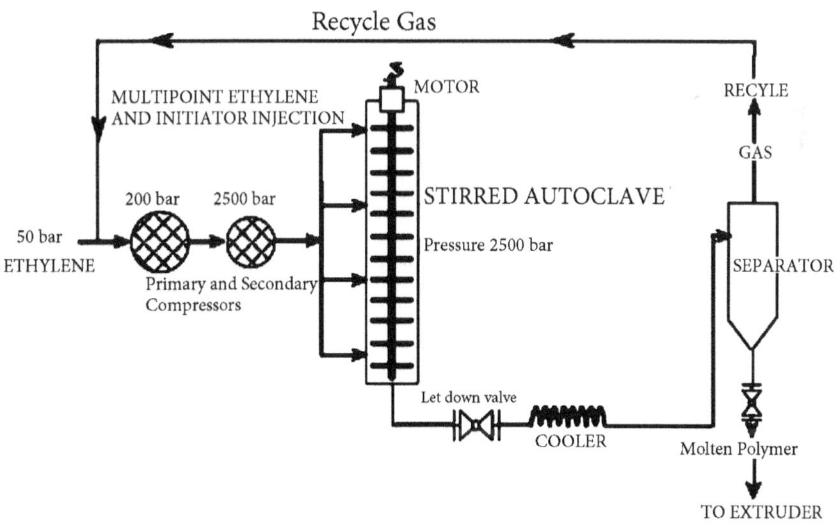

Figure 13.1

High pressure stirred autoclave

A schematic of a tubular reactor is shown in **Figure 13.2**. The tube is jacketed to allow the use of a coolant to remove exothermic heat and control the temperature and polymerization rate.

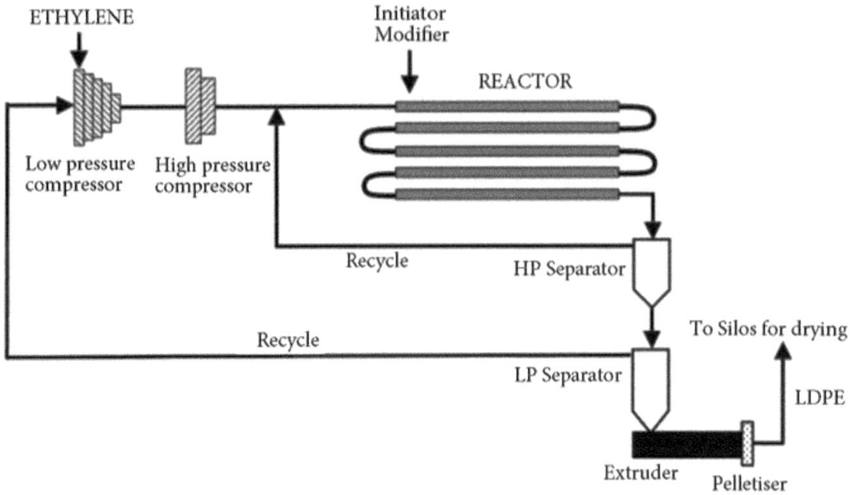

Figure 13.2

High pressure tubular reactor

The ethylene gas is delivered at around 50 bar pressure and 30 °C and compressed in 2 stages to a reaction pressure of 1500-2500 bar. The peroxide initiator solution and chain transfer agents can be injected, with auxiliary pumps, at any point along the length of the reactor. The reaction is exothermic and the ability to remove heat will determine the ethylene conversion rate and output of the reactor. The molten polymer is separated from the residual gases in the separators and ethylene is flashed-off and purified and recycled back into the process. The molten LDPE is dropped into an extruder equipped with an underwater pelletizer. The resulting LDPE pellets are then dried and blended in aerated silos. Some processes include an additional degassing stage to eliminate residual trace volatiles. Autoclaves have ethylene polymerization conversion levels of 15-18% per pass. The best tubular reactor designs can achieve ethylene conversions of 35-40%.

The high pressure process, tubular or autoclave produces high levels of branching (long and short chain) thus limiting the upper density range to around 0.932 g/cm^3. At the upper densities operating costs and ethylene polymerization rates become less attractive. The low pressure catalytic processes are the preferred routes for the production of medium density (0.935-0.940 g/cm^3) polyethylenes.

Commercial LDPEs range in melting point from 110-115 °C. Variations in operating temperature, pressure, chain transfer agent and reactor configuration will produce resins with varying degrees of long and short chain branching (SCB), molecular weights and molecular weight distributions (MWD). The SCB is made up mainly of C_2 to C_5 chain lengths. The long chain branching (LCB) is more difficult to characterize.

An advantage of the free radical initiated process is the absence of catalyst residues, particularly heavy metals, catalyst scavengers and solvents, which are present in trace amounts in the catalytic processes. The stirred autoclaves are more conducive to producing resins with wider MWD and higher levels of LCB resulting from the increased probability of back mixing in the reactor.

Figure 13.3 shows typical LDPE molecular weight distribution (MWD) curves. MWD is defined as the ratio of the weight average molecular weight (Mw) to the number average molecular weight (Mn).

$$\mathrm{MWD} = \frac{\mathrm{Mw}}{\mathrm{Mn}} \qquad (13.1)$$

The MWD is called narrow if the polymer is made up of chains close to the average length. If the chain lengths vary widely the polymer will be termed as broad or wide in MWD.

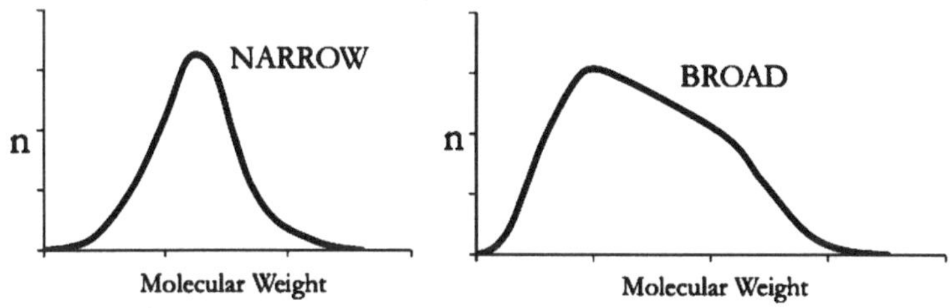

Figure 13.3

Typical MWD of LDPE

As a general rule narrow MWD LDPEs will have better opticals and flex crack resistance and better drawdown than the broad MWD grades. Resins with broad MWD tend to have higher viscoelasticity and produce hazier films. The broader the MWD the more shear sensitive is the polymer in the melt state, which will reduce torque and back pressure in extrusion machines. This phenomenon makes LDPE one of the easiest polymers to extrude at high outputs. For extrusion coating a wide MWD distribution curve with a shoulder at its high end is preferred. For tubular (blown) film applications narrower MWD are recommended.

The MWD ranges produced by the basic low density polyethylene processes are given in **Table 13.1**.

Table 13.1

Molecular weight distributions (MWD) of low density polyethylenes

Type	Process	Typical MWD
LDPE	High pressure	5-8
LLDPE (Ziegler-Natta)	Low pressure	3-4
Metallocene LLDPE	Low pressure	~2

Extrusion coating grade LDPE should be polymerized with very wide MWD and long chain branches. The best grades are produced in autoclave reactors. The main concern for the extrusion coater is to minimize neck-in and edge beading at the very high melt temperatures utilized in the process. This is best achieved with broad MWD and LCB. However, it is important to minimize the presence of very low molecular weight species as these will oxidize very rapidly leading to poorer organoleptic properties. Ideally oxidation should be formed on the longer chains and not the waxy material.

Tension thinning polymers such as LLDPE, HDPE and polypropylene (PP) have excessive neck-in and will also develop draw resonance in the air gap between the die and the chill roll nip. The longer the air gap the more pronounced these instabilities become. These materials are characterized by narrow MWD and most crucially the absence of any measurable degree of long chain branching.

13.2 High pressure ethylene copolymers

The high pressure process can tolerate comonomers incorporating heteroatoms. This allows ethylene copolymers with ester and acid functionality to be produced. The best known are listed in **Table 13.2**. To date none of the catalytic processes (Ziegler-Natta or metallocenes) are capable of handling monomers with heteroatoms.

Table 13.2

High pressure ethylene copolymers

Type	Abbrev.	% Comonomer
Ethylene-acrylic acid	EAA	2-9
Ethylene-ethyl acrylate	EEA	5-20
Ethylene-methacrylic acid	EMAA	2-9
Ethylene-methyl acrylate	EMA	5-25
Ethylene-n-butyl acrylate	EnBA	5-20
Ethylene-vinyl acetate	EVA	2-32

These copolymers like their homopolymer analogues will contain both long and short chain branching in addition to the comonomer group. Their processing characteristics will be similar to LDPE in terms of their shear and elongational viscosity. The comonomer level can be theoretically be varied from 1 to 40% by weight. The basic structure is shown schematically in **Figure 13.4**. Vinyl acetate and alkyl acrylates are shown as typical comonomers.

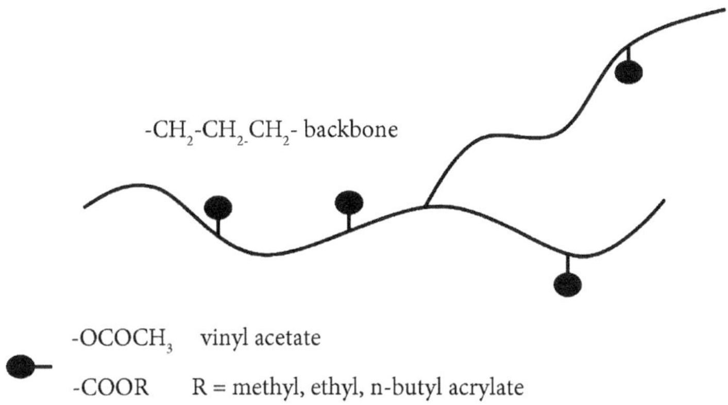

Figure 13.4

Basic copolymer structure

In **Figure 13.5**, as comonomer content is increased crystallinity is reduced, which in turn depresses melting point and stiffness. Films extruded from these copolymers have good low temperature impact strengths and excellent heat seal characteristics. A key feature is their increased polarity induced by the oxygenated comonomers. Polarity enhances adhesive properties and compatibility with a broad range of organic and inorganic materials.

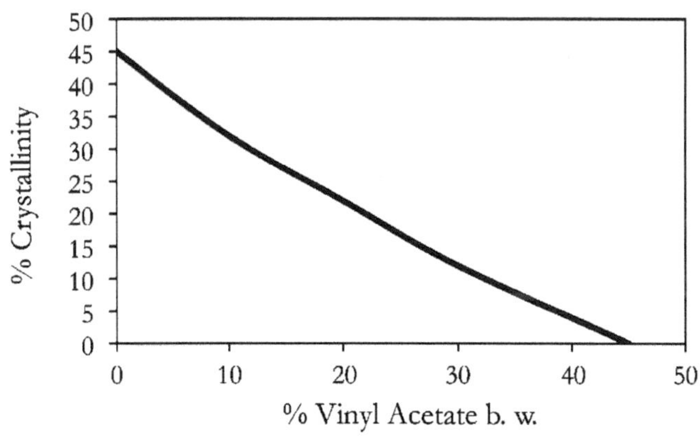

Figure 13.5

Effect of comonomer on crystallinity

13.2.1 Ethylene-vinyl acetate copolymer (EVA)

Vinyl acetate (VA) is the easiest monomer to polymerize with ethylene since both have similar reaction kinetics in the high pressure process, i.e., both monomers are inserted at the same rate and copolymerize randomly. High clarity tough films and coatings with excellent heat seal are produced.

At high VA levels the coatings tend to have tacky surfaces and consequently embossed chill rolls should be used to break up the surface and prevent chill roll sticking. In extrusion coating EVAs incorporating 16-18% VA can be coated on BOPP and OPET films. These resins have melt indexes of 12-15 dg/min. The films are corona or flame treated. Extrusion temperatures should be below 260 °C. For film processing VA levels of 3-9% are used. These films are widely used to produce lamination films particularly in blends and coextrusions with LLDPE.

EVA coated OPET films are used in heat lamination processes to provide protection and stiffness and enhance the appearance of prints, photographs, book covers, documents and other flat surfaces. The EVA used is easily heat activated. Other applications are in heat sealable lidding film for trays used in packaging. The seals can be made peelable.

Sealable barrier films have also been produced using oxide coated OPET film plus an extrusion coated EVA sealant layer. The barrier layer can be either silica (SiO_x) or aluminium (Al_xO_y) applied using a vacuum deposition process. The EVA adheres well to the very thin oxide layer.

The polar acetoxy group has a significant effect in improving compatibility with other materials particularly inorganic substances (chalk, carbon black, titanium oxide, silica, etc.) However, the (-C-O-$COCH_3$) group is also thermally labile and breaks down at temperatures >230 °C leading to the formation of acetic acid and some crosslinking. **Figure 13.6** illustrates the weak oxygen link between the ethylene chain and the acetate group. Breakdown occurs at the oxygen atom in the molecule to form acetic acid and unsaturation in the residual chain, which can then crosslink.

In contrast the acrylate in the EMA, as shown, is linked to the chain through the much stronger (-C-C-) bond and consequently this class of copolymer can tolerate temperatures above 300 °C. **Figure 13.7** shows the effect of repetitive extrusions on the degradation (melt index change) of EVA versus the more stable EMA. The EVA (2 dg/min. MI) incorporates 18% VA and the EMA (6 dg/min. MI) contains 21% methyl acrylate.

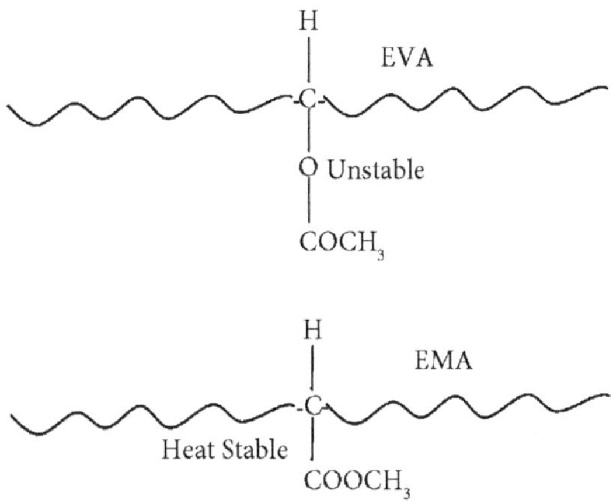

Figure 13.6

Vinyl acetate and acrylate structures compared

In **Figure 13.7**, the 'new' melt index is divided by the original value. The EVA drops in melt index as it crosslinks with each extrusion pass. The EMA shows only a slight increase in melt index after several passes at 300 °C extrusion temperature.

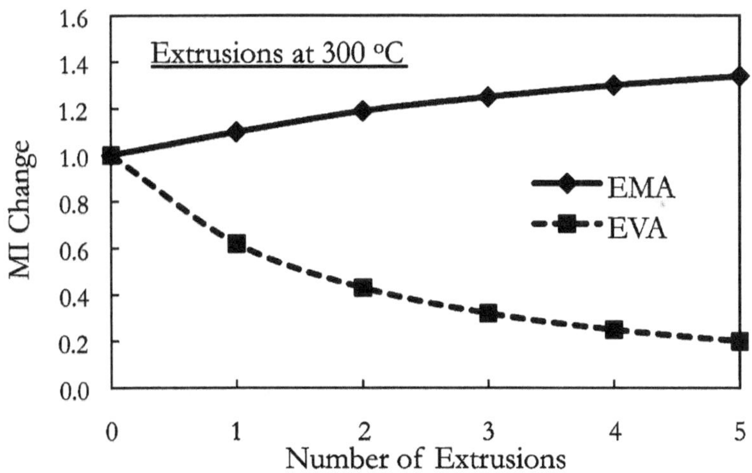

Figure 13.7

Melt index change vs. extrusion passes

13.2.2 Ethylene-alkyl ester copolymers (EAE)

EMA, EEA and EnBA are the three generic types of ethylene-alkyl esters (EAE) currently available commercially. These are more heat stable than EVAs as shown above, but are more costly and more difficult to polymerize with ethylene. Removal of trace comonomer from the product to minimize odour can be problematic.

The acrylate films develop more haze than their EVA counterparts. EMA is the most heat stable of the acrylate family and has found some applications in extrusion coating. Textile coating is believed to be an outlet for these materials. There is little evidence that the acrylate copolymers are used in lamination films.

Table 13.3 shows some extrusion coating adhesion data for EMA on three flexible packaging films. The films were corona treated to maximize adhesion. Adhesion values on the three films tested are good. The LDPE adhesion values are too low for commercial use.

Table 13.3

Comparative EMA adhesion

Material	BOPP	PVDC/cellophane	OPET
	Adhesion (N/15 mm width)		
LDPE (control)	0.2	0.6	0.5
EMA	3.0	4.1	4.0

These EMAs can be coextruded with LDPE and extrusion coated on plastic films with good results. Melt temperatures of 300-310 °C are suggested. Some manufacturers will recommend the EEA for similar applications. The alkyl ester and vinyl acetate copolymers are interchangeable for many applications particularly in coextrusion. They are not recommended for coating on aluminium foil.

Ethylene copolymers are usually produced in autoclave reactors, which results in random distribution of the comonomer. In tubular reactors it is claimed that distribution can also be controlled yielding resins with higher melting points. Tubular processes capable of producing EMAs with up to 25% comonomer have been commercialized. This technology produces a very soft material that retains some 15% crystallinity with 25% MA incorporation. The process controls the distribution of the MA in the polymer chains and the small size of the MA molecule allows for very close molecular packing. Excellent adhesion is claimed to other polyolefins.

13.2.3 Ethylene-acid copolymers

This class of copolymers are also produced by the high pressure LDPE process using either acrylic acid (AA) or methacrylic acid (MAA) as comonomer. The difference in molecular weights should be taken into account when comparing these two carboxylic acids. As an example 10 wt% AA is equivalent to 12 wt% MAA. These copolymers contain hydrogen bonds in their structure, which enhances their physical properties. The schematic in **Figure 13.8** shows the hydrogen bonds joining two polymeric chains.

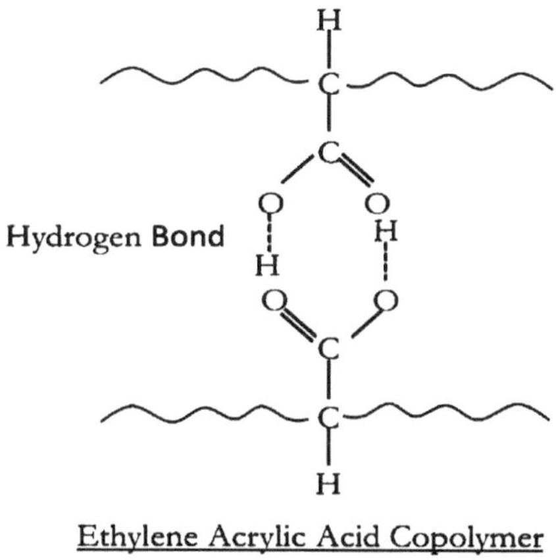

Figure 13.8

Formation of hydrogen bonds

The carboxylic acid incorporation does not normally exceed 10% by weight for products intended for film forming and extrusion coating processes. The effect of acrylic acid incorporation on stiffness (E-modulus) is shown in **Figure 13.9** and compared with vinyl acetate. The data shows that the EAA copolymer is stiffer than the EVA over the same level of comonomer addition. The presence of hydrogen bonds in the carboxylic acid copolymer counteracts to some extent the softening effect of comonomer addition.

EAA and EMAA copolymers have excellent adhesion to metal and pulp and paper substrates. The carboxylic acid is believed to form either covalent links and/or hydrogen bonds with the substrate. Oxidation is **not** required as with LDPE. The degree of hydrogen bonding is increased with acid content, which enhances mechanical properties and increases melt strength.

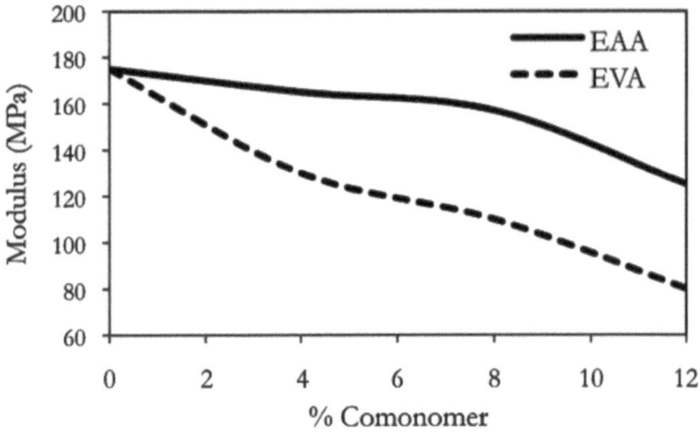

Figure 13.9

Effect of comonomer on modulus

Key properties of E-AA/MAA are:

- Excellent adhesion to polar materials (metals and paper)
- Good optics
- Toughness
- Good oil and grease resistance
- Excellent hot tack
- Sealability through contaminants
- High stress crack resistance
- High impact resistance
- High melt strength (low neck-in)

EAA copolymers designed for extrusion coating have excellent 'draw-ability' allowing very thin coatings, and the lower extrusion temperatures result in much reduced odour and off-taste. The bonds with aluminium foil have excellent resistance to alcohols, vegetable oils, and low acid foods (pH >6.4).

EAAs are the preferred materials for coating on aluminium foil bearing laminates. Acid levels of 9-12% will have superior adhesion compared with lower contents of 3-6%. However, the adhesion to polyethylene drops at high acid levels in coextrusions of these two materials.

The preferred usage of EAA is in coextrusion coating with LDPE to maximize adhesion and product resistance. Aseptic packaging of milk and juice products is a large market segment.

Other packaging applications include shampoo, toothpaste, fresheners, condiments, soup and various pouches and sachets. EAAs are also used as thermal films either as monolayers or in coextrusion with other copolymers.

13.2.4 Ionomers

Ionomers are metal salts of the carboxylic acid copolymers described above. The EAA or EMAA is neutralized in a melt process with reagents bearing either zinc or sodium cations. The reaction is as shown in **Figure 13.10**. The metal cations in the polymer matrix form ionic crosslinks, which significantly reinforce the polymer's structure. These crosslinks are reversible at high temperature thus allowing the ionomer to be processed on normal polyethylene extruders.

Being a copolymer melting point and stiffness are reduced as crystallinity is depressed by the presence of the carboxylic acid comonomer. However ionomers are unique materials since they contain some hydrogen bonds formed by residual non-neutralized acid moieties as well as the very strong ionic crosslinks that result from the presence of powerful cations. In **Figure 13.11** the increase in stiffness (E-modulus) and film impact strength as a function of neutralization (formation of ionic crosslinks) is shown. The more cation is incorporated the harder the film becomes.

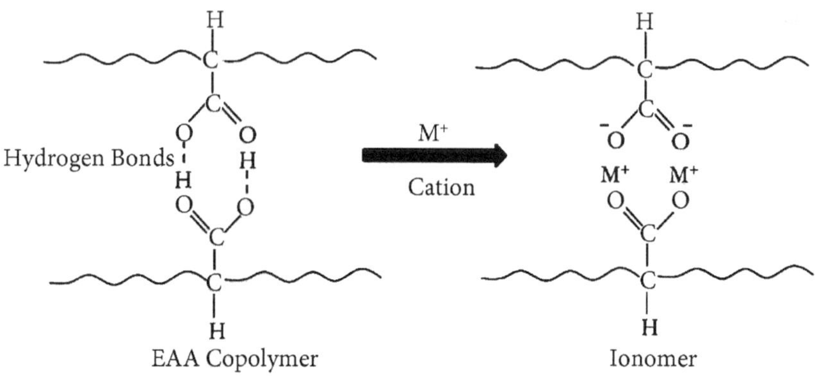

Figure 13.10

Ionomer formation from EAA

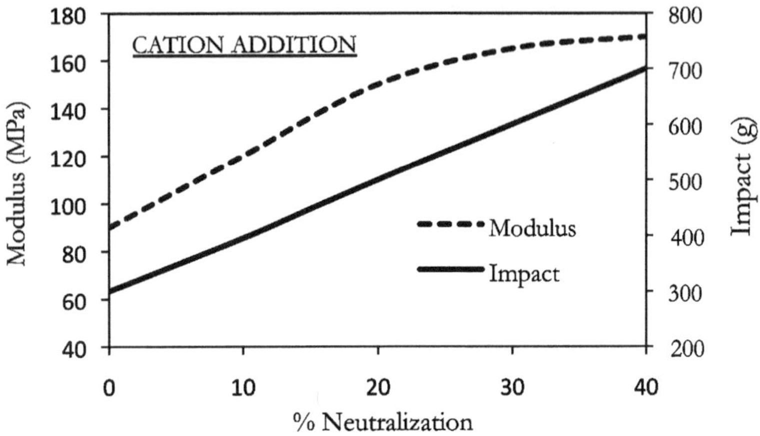

Figure 13.11

Effect of neutralization on properties

The net effect is a material that embodies the following property improvements:

- Very high melt strength: excellent thermoforming
- High hot tack
- Good oil resistance
- High clarity
- Wide sealing window
- Low sealing temperature
- High tensile strength
- High impact strength
- High-moderate stiffness

Ionomers are used in blown and cast film processes and the zinc version in extrusion coating where it has excellent adhesion to aluminium foil. Ionomers can also be coextruded with polyamide (PA) without the need for a tie-resin. Since the precursor EAA or EMAA are produced by the LDPE high pressure process ionomers have relatively wide MWD plus some long chain branching, which results in shear and elongational viscosities similar to LDPE. For extrusion coating melt indices of 4.0 to 12.0 dg/min. are used. Ionomers absorb water and must be contained in moisture proof packaging. The sodium grades are the more sensitive to moisture. Bags or boxes should only be opened when the material is ready for use.

An interesting feature of ionomers is that they combine very high impact strength with high stiffness. Ionomers are comparable to MDPE (medium density) in their rigidity, but are much tougher. **Figure 13.11** illustrates this effect.

New ionomer products have been developed, which exploit the ability to combine high stiffness with low sealing temperature. These materials are based on ethylene copolymers with up to 20% carboxylic acid, which is then 70-75% neutralized. This results in a very high degree of ionic crosslinks in a low crystallinity matrix. These grades can be sealed at temperatures as low as 80 °C and have a flexural modulus above 500 MPa compared to 180 MPa for LDPE.

Blends of ionomers with EVOH and polyamides have been studied as a way to reduce the barrier resin's high cost and maintain low permeability to gases. Various papers and patents teach methods of forming blends where the barrier polymer forms a laminar phase in the structure. The theory is based on the interaction of polymers containing hydrogen bonds, e.g. EVOH and ionomers, which combine through intermolecular interaction to achieve high barrier properties.

13.3 Linear low density polyethylene (LLDPE)

LLDPE was invented by DuPont in the late 1950s. However, it was not until the early 1980s that the technology was fully exploited on a global basis. The low pressure co-ordination catalytic processes control density by the incorporation of 4-10% of α-olefin. The catalysts are usually of the Ziegler-Natta (Z-N) type. The comonomers in commercial use are shown in **Figure 13.12**. The comonomer introduces short chain branching (SCB) onto the main polyethylene backbone,

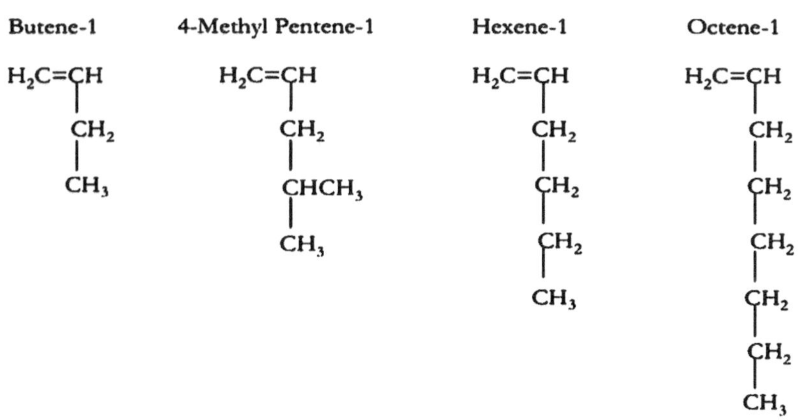

Figure 13.12

Alpha-olefins used in LLDPE production

which inhibits the close packing of the polymeric chains thus reducing density (crystallinity). Density is determined by the amount of comonomer inserted. Melting points are in the region of 120-125 °C, and higher than LDPE (~110 °C) of equivalent density. It is generally accepted that the best balance of properties is achieved with n-octene.

The low pressure processes do **not** form any measurable amounts of long chain branching (LCB). One subtle but important difference between LDPE and LLDPE is that the composition distribution (CD) is heterogeneous in LLDPE and relatively homogeneous in LDPE, particularly from tubular reactors. The Ziegler-Natta catalysts used have multiple active sites, which leads to a broad composition distribution (CD). These aspects of composition heterogeneity are discussed in greater detail in section 13.16.

Commercial LLDPEs are usually produced by either gas phase or solution processes. More recently slurry loop reactors used for HDPE, have been successfully adapted for LLDPE production. All have their merits and are capable of producing a range of melt indices and densities suitable for film applications. In extrusion coating LLDPEs should be blended with LDPE to achieve acceptable melt stability and reduce pressure in the extruder and die.

Figure 13.13 is a schematic of a fluidized gas phase process. The gas phase process operates at a pressure of 20 bar and a temperature of 80-100 °C. Density

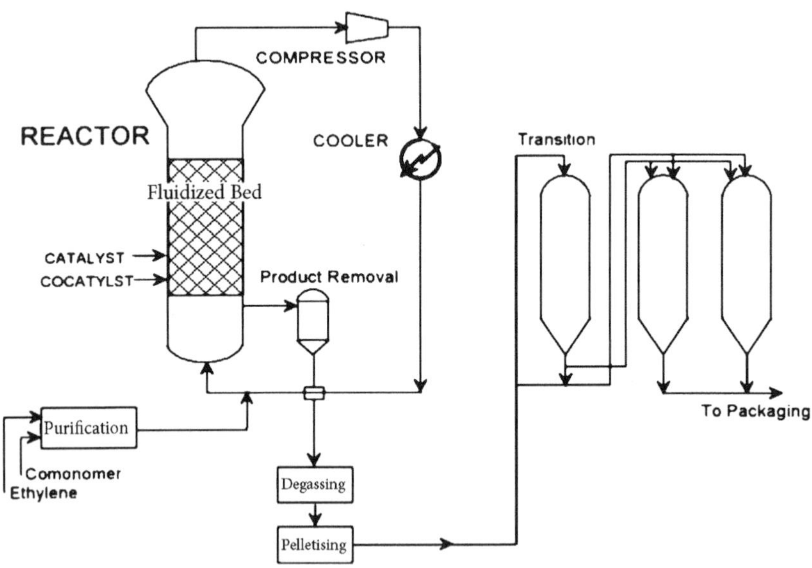

Figure 13.13

LLDPE fluidized gas phase process

is controlled by the ratio of ethylene to comonomer. The bed of growing polymer particles is agitated either by fluidizing the bed or by stirring. The inert medium in the reactor is usually nitrogen, and hydrogen is used to control molecular weight (melt index). A fluidized reactor bed consists of organo-metallic catalyst particles that are 'fluidized' by the flow of ethylene gas. The catalyst particles are suspended in the ethylene gas, which is pumped from the bottom of the reactor bed to the top. Fresh catalyst and cocatalyst is pumped with the ethylene stream into the reactor as the polymerization progresses. The catalyst is incorporated with the polymer as the growing molecules remain linked to the catalyst particle from which they were produced. Exothermic heat is released from the polymerization. Heat is removed by cooling unreacted ethylene gas coming off the top of the reactor and recycling the cooled gas back into the system. Reactor capacity is to a large extent determined by the ability of the circulating gas to remove heat.

The conversion of ethylene is low for a single pass through the reactor. Unreacted ethylene is removed off the top of the reactor, where it is expanded and decompressed to separate the catalyst and low molecular weight polymer from the gas. After purification, the ethylene gas is then recompressed and recycled back into the reactor. Granular polyethylene is gradually removed from the bottom of the reactor as soon as reasonable conversions have been achieved. Typically, a residence time of 3 to 5 hours results in a 97% conversion of ethylene. Technologies such as Univation's condensed mode technology have been developed to increase reactor capacity.

The low temperature and pressure utilized minimize energy costs. Narrow MWD (3-4) resins are produced although the process is adaptable to different catalyst and multiple reactor configurations to produce broader MWD. The cost penalties involved in interchanging incompatible catalysts can be high and only justified for extended production campaigns. HDPE may also be produced in these reactors. Gas phase reactors of up to 450 kt/yr. capacity have been constructed.

The inability to gain wide market acceptance for LLDPE granules straight out of the reactor has compromised the cost competitiveness of the gas phase process versus other technologies. Large compounding extruders are required to convert the granules into free flowing pellets, which are preferred by processors. The addition of large capacity homogenization extruders adds both to the investment and operating costs of the process.

The replacement of n-butene by n-hexene as commoner will reduce reactor output, making these copolymers more expensive to manufacture. The higher α-olefins copolymers, however, do have superior physical properties, which justifies their higher cost in some applications.

Table 13.4 compares a high pressure LDPE to a gas phase LLDPE (butene). The tests were carried out on 37 μm film produced at 2 : 1 blow-up ratio.

Table 13.4
LLDPE compared to LDPE

Property	Unit	LLDPE	LDPE
Melt index	dg/min	1.0	1.0
Density	g/cm^3	0.918	0.918
Dart impact	g	150	90
Puncture	J/mm	68	27
Elmendorf tear MD/TD	g/μm	3.5/9.2	6.3/3.7
Tensile strength MD/TD	MPa	38/31	17/14

At equivalent melt index and density the LLDPE is superior in physical properties except for MD tear strength. Puncture resistance and tensile strength are significantly higher with the linear polymer.

As the name implies the solution process produces polymer in a solvent, which is then flashed off. The process runs at 140-150 °C and 40-80 bar to ensure the polymer remains soluble in the solvent during polymerization. Productivity increases with increasing reaction temperature. The process operates efficiently with a wide range of comonomers from butene to octene. High comonomer ratios are accommodated making it possible to reduce density below 0.915 g/cm^3. One advantage of the solution process is its greater flexibility in grade changes compared to gas phase processes. In comparing the efficiency of competing polymerization processes the cost of grade changes and reactor flexibility should be factored into the final audit. A schematic of a typical solution process is shown in **Figure 13.14.**

The process and catalyst options to produce linear polyethylenes are being continually extended by all the major polyolefin producers trying to position themselves in this intensely competitive industry. As an example the well established slurry loop reactor process for HDPE has been adapted for the manufacture of LLDPE.

The superior physical properties of LLDPE films compared to conventional LDPE were recognised at their early stages of development. In film applications their improved heat sealability, higher tensile strength and low temperature toughness all encouraged interest in their use.

DuPont was probably the first to recognise these attributes in their Sclair LLDPE resins and films. Sclair resins, introduced in the early 1970s, were produced by a solution process. These resins were targeted at speciality applications such

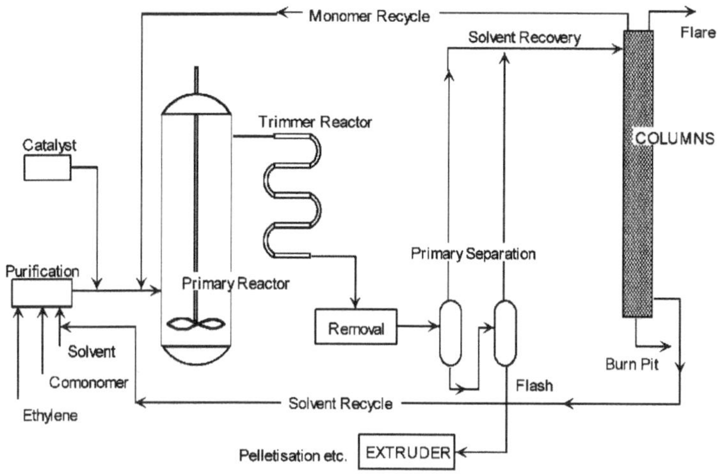

Figure 13.14

Typical LLDPE solution process

as lamination films and coextruded milk pouches. The key benefits of LLDPE were a reduction in milk pouch leakers and improved hot tack and seal strength in lamination films. In Canada 60% of fresh milk is packaged in polyethylene pouches. India is another large volume user of pouches for milk distribution. The LLDPE could also be biaxially oriented in double bubble processes to produce high performance shrink films. These have balanced shrink properties, are highly transparent and easily heat sealable.

By reducing the level of comonomer, densities can be boosted to ranges >0.930 g/cm^3. These materials are defined as medium density polyethylenes (MDPE) and are an intermediate class of products positioned between LLDPE and HDPE.

In **Figure 13.15** typical blown film properties of LLDPEs made with three α-olefin comonomers are compared subjectively. The rule-of-thumb is that the larger the comonomer molecule the better the properties of the film. This is attributed to the probability of more chain entanglements as the side chain increases in length. However, making simple product to product comparisons can be misleading since other structural variables such as morphology, molecular weight and composition distributions must be taken into account. All of these variables will have some effect on both melt flow and solid state properties, which may mask the contribution from the specific comonomer utilized. However, as a basic rule the higher α-olefin LLDPE grades are superior in ultimate elongation, puncture resistance, tear strength and dart impact to the n-butene grades at equivalent melt index and density. This is reflected in the pricing structures for the materials.

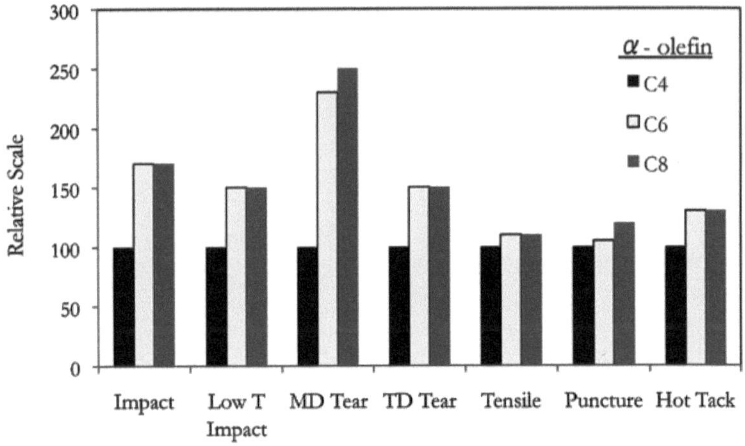

Figure 13.15

Comparative effect of α-olefin selection on properties

13.4 Melt viscosity of LD/LLDPE

An important difference between LLDPE and LDPE is in their relative ease of extrusion. Because of its narrower molecular weight distribution, LLDPE develops a higher shear viscosity than LDPE. This is demonstrated in **Figure 13.16** where the shear viscosities of ~1 dg/min. melt index LLDPE and LDPE are compared at 230 °C. A metallocene grade LLDPE is also included to highlight the effects of a further narrowing of molecular weight distribution (MWD).

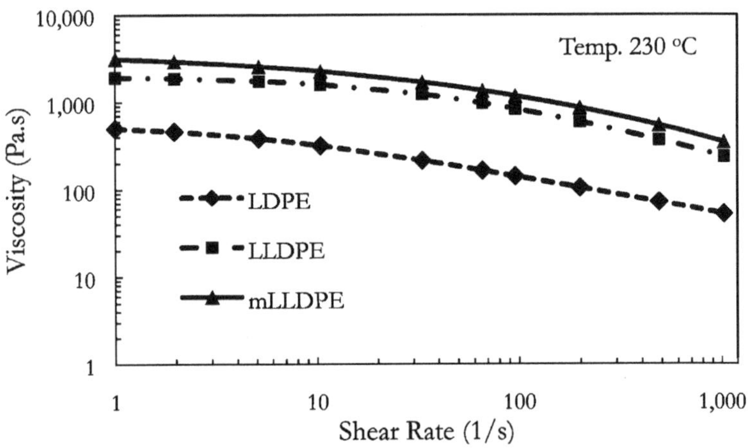

Figure 13.16

Shear viscosity of LDPE and LLDPE

A simple calculation comparing the flow through an annular channel typical of the lips of a blown film die illustrates this crucial difference between LDPE and LLDPE. The melt index is ~1 dg/min. for both resins. The annular die dimensions and inputs are shown in **Table 13.5**.

Table 13.6 summarises the flow behaviour of the two polymers through this simple geometry. The calculation is based on comparative capillary viscosity (Carreau model) data.

Table 13.5

Die dimensions and inputs

OD diameter	300 mm
ID diameter	303 mm
Land	10 mm
Die gap	1.5 mm
Output	200 kg/hr.
Temp.	220 °C

Table 13.6

Comparative flow of LDPE and LLDPE in annular die

Flow	LDPE (MI 0.8 dg/min)	LLDPE (MI 1.0 dg/min)
Mass flow rate (kg/hr.)	200	
Melt temp. (°C)	220	
ΔP (MPa)	1.0	3.2
Shear stress (kPa)	75.3	238

The pressure drop (ΔP) is threefold higher for the LLDPE, which is reflected in shear stresses at the walls of 238 kPa versus 75 kPa for the LDPE. This excessive shear stress will result in severe sharkskin. Widening the die gap is the easiest remedy to eliminate surface melt fracture. However, this solution will result in unbalanced orientation in the film and poorer thickness distribution from the die. Processors have developed compromise methods, such as blending with LDPE, the use of polymer processing aids (PPA) and boron or titanium nitride coated die lips to alleviate sharkskin formation.

PE suppliers and technology licensors have been active in trying to overcome these processing limitations by attempting to polymerize broad MWD LLDPE in dual reactor processes. These resins are claimed to have the processability and properties of LD/LLDPE blends. So far none have succeeded to completely mimic LDPE.

13.5 High density polyethylene (HDPE)

The types of HDPE used in blending with LDPE for extrusion coating are usually produced in solution processes using Z-N catalysts. These have narrow MWDs and high melt index (8-10 dg/min.) to enable good drawdown and avoid excessive pressures in the coating extruders. However, the highest recent growth rate in HDPE usage has been in low gauge tubular film applications.

The development of the long neck blown film extrusion process with high blow-up ratio has made it possible to produce high strength low gauge HDPE films, which compete with LDPE, LLDPE and paper in commodity bag and bin liner markets. For the process to work effectively, HDPEs of high melt strength and toughness are required. This is achieved by the polymerization of high molecular weight (HMW) resin for strength, combined with very broad MWDs for processability. Splittiness in thin films is avoided by balancing the MD/TD orientation during the tubular extrusion process.

Recent estimates show that around 26% of HDPE production is used in film applications. This is due to the following distinctive advantages of HDPE:

- High strength to thickness ratio, i.e. reduced cost
- Improvements in extrusion equipment (e.g. grooved barrel feed sections) to allow wider web production, better thickness control and higher output
- Improved grades, particularly bimodal HMW-HDPE
- Resource reduction and lower carbon footprint
- Paper-like films with easy openability

High density polyethylenes are produced by two processes. These are based on either Ziegler-Natta (Z-N) catalysts or the chromium oxide catalytic process developed by Phillips Petroleum.

For the best combination of processability and strength, the Z-N catalyst multiple stirred tank slurry processes have been very successful in the manufacture of HMW-HDPE film grade resins. In slurry processes the polymer separates from the carrier as fine particles, which facilitates the production of HMW polymer. A schematic of a stirred tank slurry process is shown in **Figure 13.17**.

2-stage series reactors will synthesise very wide MWD (>15: 1) resins. In the first reactor the low MW fraction is produced and the active particles are transferred into the second reactor to polymerize the high molecular weight component. A broad range of MWDs can be produced with this stirred tank slurry technology. The reactors are fed with the monomers, hexane diluent, catalyst and hydrogen to

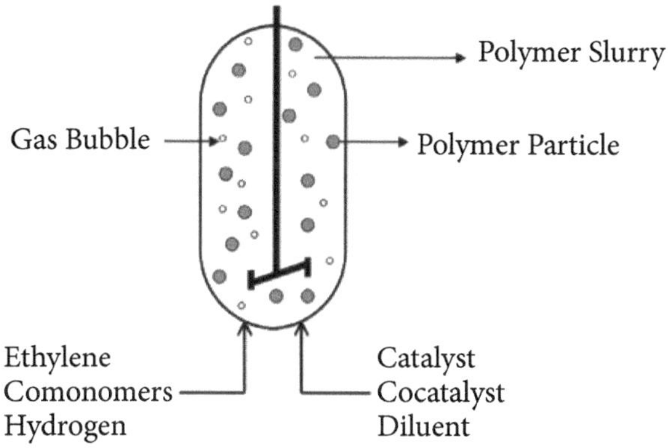

Figure 13.17

Stirred tank slurry process

control polymerization. Purified hexane is fed from the recycle train. The reaction is exothermic and the concentration of the comonomer in the second reactor is crucial in establishing the molecular distribution. The suspension exiting the reactor is fed to a centrifuge to separate the solids from the diluent. The solids are fed to a fluidized bed dryer and the liquids recycled. The dried polymer is dumped into an extruder, blended with stabilizers and pumped to a pelletizer.

The supported chromium (Cr) oxide catalyst slurry process (Phillips) produces MWD of approximately 12 : 1, in single stage slurry loop reactors (**Figure 13.18**).

These HDPEs are widely used in blow moulding, but less popular in blown film applications. However, new generation grades using this slurry process have improved significantly in recent years and are claimed by some manufacturers to be competitive in many film applications.

As the ethylene comes into contact with the catalyst the reaction is initiated and polymer is formed. The polymer is suspended in a light hydrocarbon (isobutane) diluent and never goes into solution. This enables high molecular weight polymer to be produced.

The loop reactor is essentially a pipe within a pipe which doubles as a heat exchanger. **Figure 13.19** illustrates this design where the cooling jacket (the outer pipe) is claimed to surround some 90% of the reactor surface. This results in a very efficient heat transfer mechanism achieving very high polymerization rates. The heat of reaction is first absorbed by the diluent and then transferred to the walls of the cooling jacket where water is flowing in counter current. This heat exchange process maintains a very steady reactor temperature ensuring ease of operation and product uniformity.

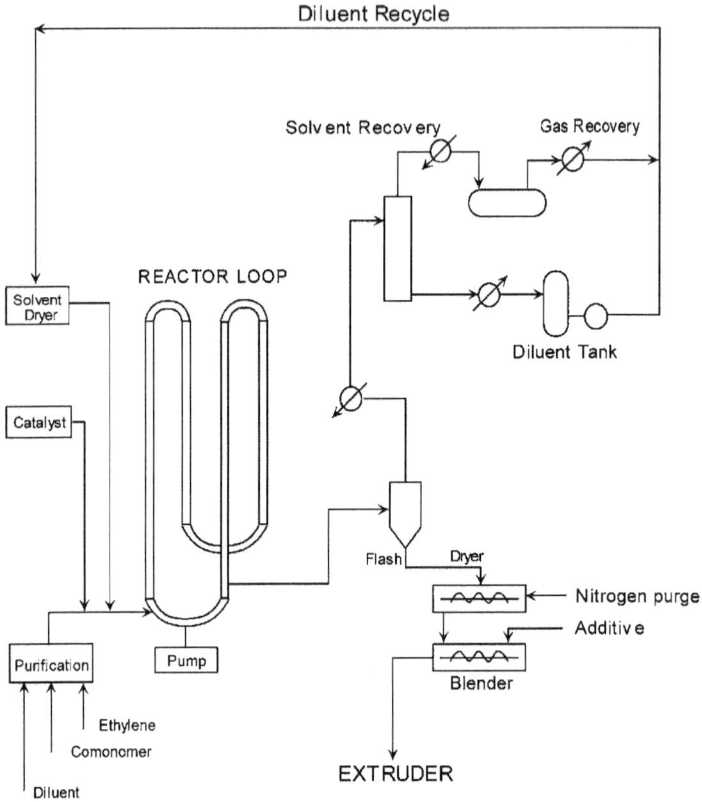

Figure 13.18

Slurry loop reactor

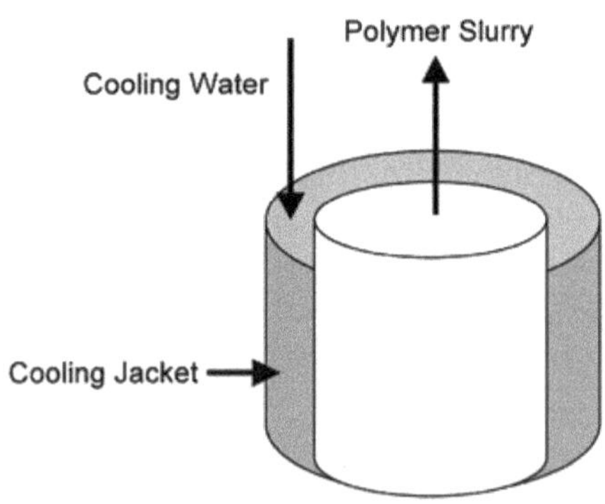

Figure 13.19

Slurry loop reactor section

The loop is usually designed with four legs with a surface area of 70 m^2 producing 165 kt/yr. of resin. Capacity can be increased with six legs providing a surface area of 100 m^2 and the output is increased to 200 kt/yr.

HDPE resins for blown film applications are divided into various molecular weight categories shown in **Table 13.7**. Melting points range from 130-133 °C.

Typical molecular weight distributions for HDPE resins used in blown film processes are 12-20 for MMW grades, and 20-40 for broad/bimodal resins.

Table 13.7

HDPE molecular weights

Type	Molecular weight (MW)
HDPE general purpose	<180,000
Medium molecular weight (MMW)	180,000-280,000
High molecular weight (HMW)	300,000-400,000

These MWD values are appreciably higher than those normally associated with LLDPE and LDPE. A total absence of gels and other hard particles in the polymer is a critical requirement for thin film extrusion. All HDPE products intended for blown film extrusion must be clean and gel free to avoid holes and breaks in the bubble.

The basic broad MWD resins produce films with acceptable toughness at a thickness of 25 μm and above, but lack sufficient strength at lower gauges. Using advanced Z-N catalysts bimodal HDPE grades have been developed as a means of maximizing strength, stiffness and drawdown with the objective of allowing down-gauging to 12 μm and below. The intent was to open up large volume opportunities for HDPE films to replace LL/LDPE and kraft paper in bags, sacks and liner markets. Another important application was in pressurized pipes.

Bimodal HDPE resins are produced with reactors coupled in series using different conditions in each vessel. The polymer fractions in each reactor will have their own MWD. The final MWD consists of the different fractions superimposed on each other.

In **Figure 13.20** the molecular weight distribution of an idealized HDPE resin will have two distinct elements controlled by the catalyst used in the polymerization. The narrow distribution HMW component contributes strength and the broad distribution lower MW element contributes to processability. A bimodal structure, as shown in **Figure 13.20**, produced by proprietary Z-N catalyst formulations and series reactors combines strength attributes with

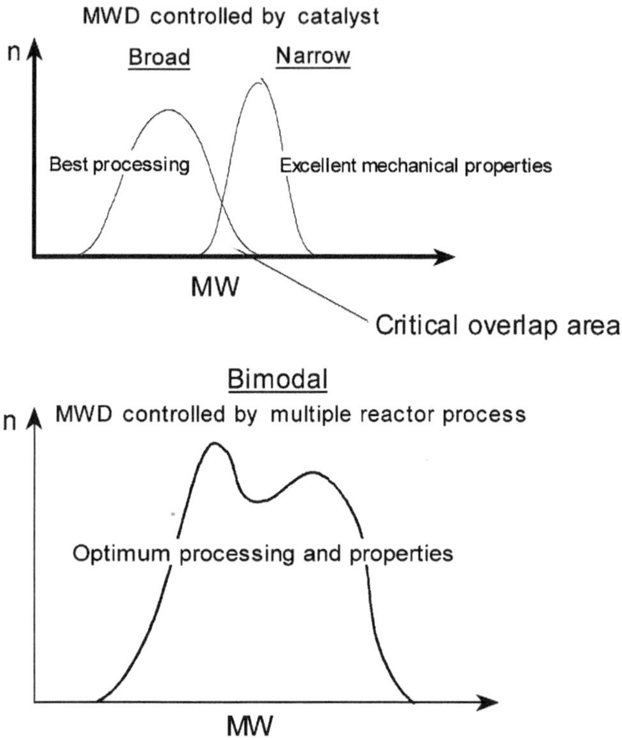

Figure 13.20

Bimodal HMW-HDPE

good processing. Densities of bimodal grades are usually between 0.945 to 0.956 g/cm^3. The MWD and SCB is more easily controlled by cascade (series) polymerization units using one catalyst. The MWD will depend on the MW and polymerization ratio in each reactor. The concentration and location of the α-olefin comonomer incorporation is critical. The overlap between the two MWD curves is important in optimizing the balance between processability and properties and the formation of gels in thin films.

Bimodal HDPE is less sensitive to MD orientation and as a consequence produces more balanced films at any given blow-up ratio. This partly explains the very high impact strengths achieved with thin films. Some basic characteristics of unimodal and bimodal HMW-HDPE grades prepared with two catalysts are summarized in **Table 13.8**.

The flow curves of a bimodal HMW-HDPE are compared to a conventional (unimodal) HDPE and LLDPE in **Figure 13.21**. The unimodal resin is produced with chromium catalyst in a slurry loop process and the LLDPE (octene) in a solution process with Z-N catalyst.

Table 13.8

Unimodal and bimodal HDPE

HMW-HDPE	Unimodal	Bimodal
Catalyst type	Cr	Ti (Z-N)
M_w/M_n	10-20	20-40
M_w	<250,000	>300,000
% Crystallinity	60-75	60-75
% CHMS (conc. HMW species >500,000)	~15	>20
% CLMS (conc. LMW species <1000)	~1	<1

The curves show that in the typical film processing range of 50-500 s^{-1} shear rate the bimodal HDPE has a lower viscosity than either the HDPE (0.1 dg/min. MI) or the octene LLDPE (1.0 dg/min. MI). The very wide MWD will be effective in reducing pressure and shear stress in extruders and film dies.

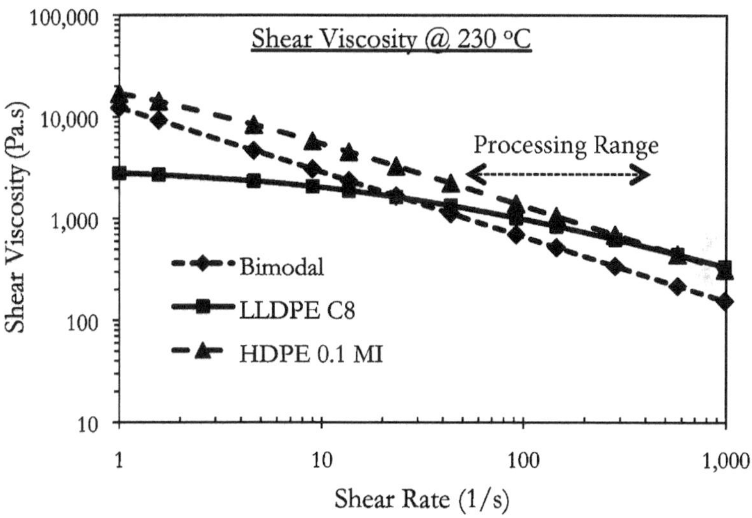

Figure 13.21

Comparative shear viscosity data

This data was used to estimate the shear stress developed in a die of 250 mm diameter and a land length at the lips of 5 mm. Mass flow rate was 180 kg/hr. at 230 °C. **Figure 13.22** summarizes the results at three die gaps.

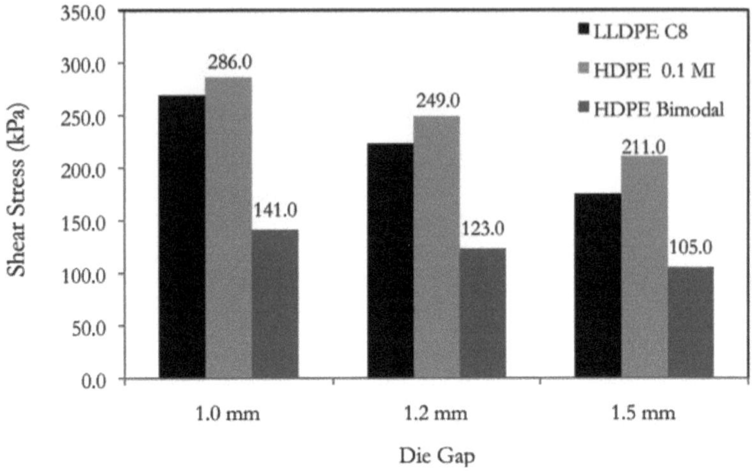

Figure 13.22

Shear stress in 250 mm tubular die

Assuming a sharkskin threshold of 140 kPa the bimodal HDPE can probably be run with a die gap of 1.0-1.2 mm. The other two resins will require die gaps well above 1.5 mm.

Film property comparisons of unimodal and bimodal HMW-HDPE resins are shown in **Table 13.9**. The improved impact strength and tear resistance of the bimodal polymer are clearly illustrated. The availability of bimodal HMW-HDPE has made it possible to down-gauge grocery bags from 20-25 μm to 12-15 μm. Similar down-gauging is possible for refuse sacks and bin liners.

Table 13.9

Comparison of bimodal and unimodal HMW-HDPE

Property (at BUR 4 : 1)	Bimodal	Unimodal
MFR5[1] (dg/min.)	0.31	0.4
Density (g/cm³)	0.944	0.948
Dart impact strength (g/12 μm)	320	140
Dart impact strength (g/25 μm)	425	225
Tear (g/25 μm) MD/TD	32/98	19/46
[1] *melt flow rate at 5 kg load*		

More recently bimodal grades with densities as high as 0.967 g/cm³ have been commercialized using single site catalysts. These resins have very high stiffness and exceptional moisture barrier resistance. They are normally coextruded with other polyethylenes.

Bimodal HDPE has excellent compatibility with other polyolefins including polypropylene. This improved compatibility is an advantage when blending with recyclates based on other polyolefins. There have been many claims made regarding the superior results obtained by blending up to 50% recyclate with bimodal HMW-HDPE compared to conventional resins.

13.6 Polyolefin blends for extrusion coating

The most practical method available to utilize LLDPE, HDPE and PP in extrusion coating is by blending with highly branched LDPE. Blending of polyolefins generally has an additive effect and is not synergistic. The blend properties will depend on the ratios of the components and will invariably be a dilution of the individual physical, thermal and flow properties. Branched LDPE differs significantly from linear PEs and PP in their elongational flows outside the die, i.e. strain hardening versus strain thinning as discussed earlier in section 1.6.

Gravimetric feeders are recommended for blending. Ideally each feeder is placed above the extruder throat. This eliminates the possibility of desegregation. The downside of this system is the high cost of the feeders and the availability of space above the extruder. For 2-component blends this technique is recommended. If pellets of similar size are used, the simple tumble blender can also be considered. It is important to ensure accurate weighing of the pellets prior to mixing.

Blending with LDPE will reduce neck-in and reduce shear viscosity in the extruder and die. The blends should be LDPE rich to minimize melt instability in the air gap. PP is not easily miscible with LDPE and requires an additional melt homogenization step.

Blends of HDPE with LDPE are prepared by dry blending pellets and feeding directly into the hopper of the coating extruder. HDPE is usually added to LDPE to reduce moisture vapour transmission rate and to increase stiffness and hardness. HDPE is used in photographic paper coating to off-set curling of the two-side coated paper. For this application the HDPE content in the blend is usually 30-35% b/w. Other applications for HDPE blends are as an anchor layer on paper for silicone coating used as release liners.

High melt flow rate HDPE resins are used and very high density is recommended to maximize the benefits of this highly crystalline polymer. These grades will have the following properties:

- Melt flow rate: 20 dg/min. at 230 °C
- Melt index: 8.0 dg/min.

- Density: 0.962 g/cm^3
- Narrow MWD

HDPE is easily dry blended with LDPE coating grades of 4-8 dg/min. melt index.

There are patents that recommend the use of bimodal HDPE resins blended with LDPE. These HDPEs are claimed to develop less neck-in and improved melt stability. The result is higher melt strength with acceptable melt flow properties[1].

HDPE has excellent chill roll release properties at high line speeds. This characteristic can be exploited either by blending or coextrusion with LDPE. HDPE can provide the following property improvements:

- Low MVTR
- High stiffness
- Improved scuff resistance
- Improved release properties
- Low gel levels
- Grease resistance

LLDPE has better flex crack resistance and forms stronger heat seals than LDPE. This is particularly true of the higher α-olefin grades. These improved attributes are useful in some packaging applications. For the best combination of properties, metallocene LLDPEs are recommended over the Ziegler-Natta polymers.

Extrusion coating machines are normally fitted with long LD ratio screws, have high compression ratios and shallow flighted channels and will, therefore, develop excessive torque and pressure with LLDPE. This makes it difficult to run this polymer at high outputs on such machines.

The performance of an LLDPE (octene) was compared to a coating grade LDPE in a PC simulated 114 mm diameter extrusion coating screw. This was a 32 : 1 LD ratio design with channel depths of 3.5 mm in the metering section. The screw was run at 100 rpm with the following barrel temperature settings:

30/240/280/300/320/320 °C.

Figure 13.23 plots the temperature profile in the extruder for the two materials. The LLDPE temperature rises rapidly and overshoots to 330 °C. The LDPE is delivered at the required coating temperature of 315 °C. In **Figure 13.24**, the

1 Laiho, E., Vaehaelae, M. and Sainio, M., USPTO patent application 20080261064

LLDPE pressure peaks to 24 MPa and the LDPE to 8.6 MPa. The calculated screw power consumption shows that the linear polymer draws more than double the energy (65.1 kW) compared with that consumed by the LDPE (24.5 kW).

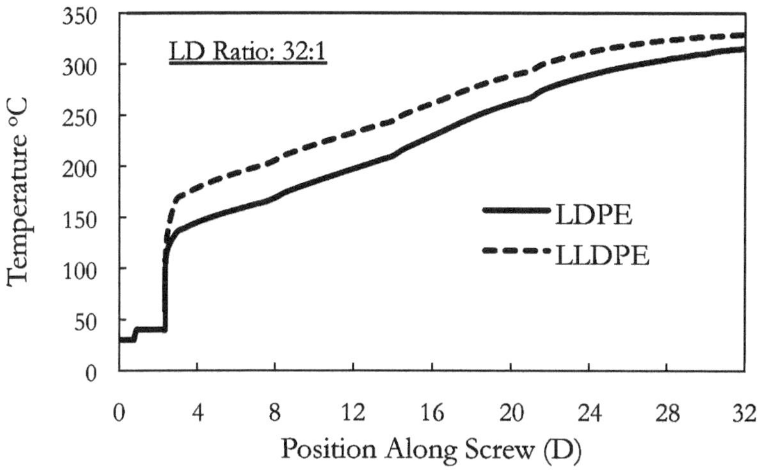

Figure 13.23

Temperature profile in extruder

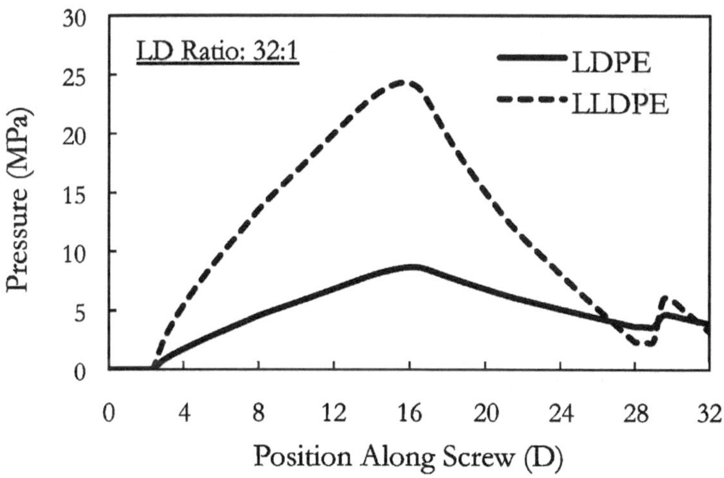

Figure 13.24

Pressure profile in extruder

Blending with LDPE is the best method to attenuate these excessive pressures and temperatures. This is best carried out with polymers of similar melt index. Adding 40% LLDPE will double the neck-in. In a typical example neck-in will increase from 30 mm to approximately 60 mm at a temperature of 315 °C. The motor load may also increase by some 40-50% depending on the blend ratio. **Figure 13.25** shows the effect on shear viscosity of blending two ratios of LDPE with LLDPE. The best results are obtained with linear **lean** blends (<50% LLDPE).

As a general recommendation, if there is a justifiable need to use LLDPE it is recommended to use a metallocene grade, which will maximize the benefits from the linear polymer. The use of very low density materials (density <0.915 g/cm^3) will increase flexibility and significantly reduce sealing temperature. High melt index grades of mLLDPE should be used to maximize drawdown and reduce pressure.

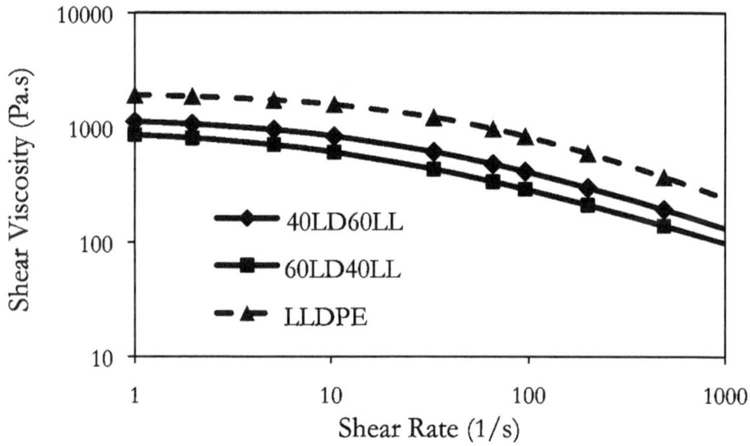

Figure 13.25

Blending LDPE with LLDPE

13.7 Metallocene-catalyzed PE

Over the past fifty years or more many new organometallic complexes have been investigated as catalysts for polymerization processes. The metallocenes as a class of compounds have been the most actively pursued by polymer producers. The term is used to describe any complex with a metal centre and cyclopentadienyl ligands surrounding it.

The research has focused on the following:

- The discovery of new active catalyst complexes
- Understanding the relationship between catalyst structure and polymer properties
- Formulation of supported single site catalysts
- Fine tuning catalysts and processes for compatibility

The differences between Ziegler-Natta catalysts and metallocenes are contrasted below.

Ziegler-Natta catalysts:

- Several metal sites results in less control of polymer branching
- Monomer insertion occurs at the **end** of the growing chain
- Changing the metal centre is ineffective

Metallocenes:

- Single metal site allows for more control of branching and molecular weight distribution
- Insertion of monomers occurs between the metal and the growing chain of polymer
- Versatility of metallocenes with many variations, (e.g. the presence of bridging atoms)

Advantages of metallocenes:

- Versatility of metallocenes for a greater range of polyolefin production
- Polymers with more favourable properties i.e. lower melting points, better optical characteristics, better heat stability and mechanical properties and lower extractables
- Increased control of side branching in polyethylene
- Yield is on the scale of a thousand times that of traditional catalysts

Disadvantages of metallocenes:

- A high concentration of aluminium is needed for the MAO (methylaluminoxane) co-catalyst. This is unacceptable in terms of the amount of residues in the polymer and the cost of aluminium. Transition metal to aluminium ratio in Ziegler-Natta catalysts is: 1:50-200. Metal to aluminium ratio in metallocene catalysts is 1 : 1000-15000.

- Relatively high cost.
- MAO is one of the elements which present a challenge to the cost reduction of metallocene catalysts. MAO is very expensive, unstable, difficult to characterise, forms gels and requires high aluminium to transition metal ratio as mentioned above.
- The alternate cocatalysts organoboranes and borates also present some challenges. These are also costly, they are not universally effective, are more difficult to support than MAO and have poor solubility in aliphatic hydrocarbons.

Figure 13.26(a) and **(b)** are simplified visualizations of a conventional Ziegler-Natta multi-site catalyzed polymerization and a metallocene single site synthesis.

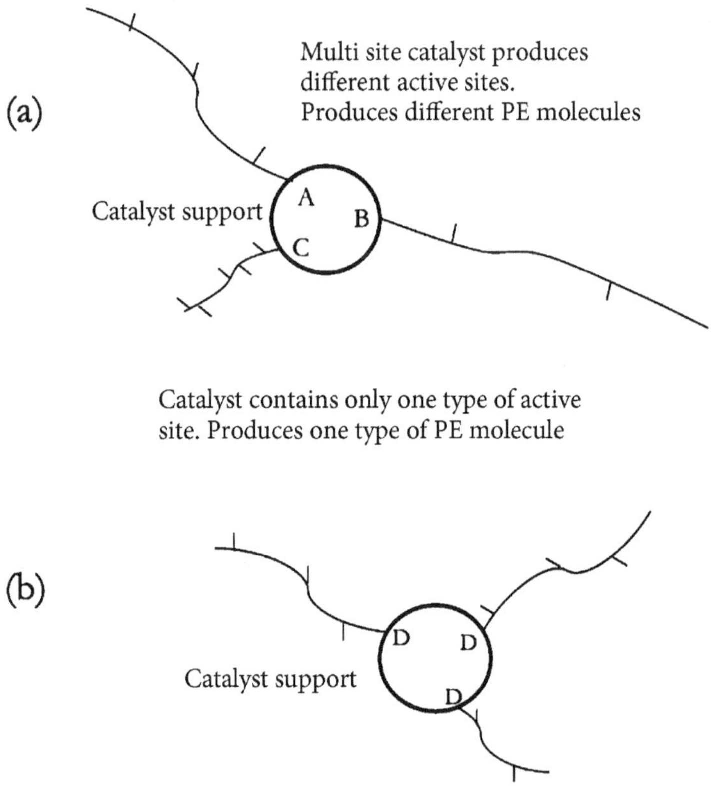

Figure 13.26

Comparison of multi-site and single-site catalyst

The conventional catalyst produces a heterogeneous polymer whereas the metallocene synthesizes a more homogenous material, i.e., similar molecular species from a single site. The molecular weight distributions (MWD) are typically 3-4 : 1 for the Z-N catalysts and 2 : 1 for the metallocenes. The comonomers used in commercial metallocene polyethylenes are usually n-hexene and n-octene. Metallocene PEs offer the following advantages:

- Low extractables particularly significant at low densities
- Improved physical properties
- Improved optical properties
- Outstanding heat seal and hot tack

The metallocene ethylene copolymers can be divided into the groupings shown in **Table 13.10**. The polyolefin elastomer products are targeted primarily at the thermoplastic rubber market.

Table 13.10

Types of metallocene PE

Polymer	Density	Value
mLLDPE (metallocene)	> 0.915 g/cm^3	Competitive with LLDPE
mMDPE (medium density)	0.930-0.936 g/cm^3	Adds stiffness
Polyolefin plastomers	0.895-0.915 g/cm^3	Comonomer <20%
Polyolefin elastomers	0.865-0.895 g/cm^3	Comonomer > 20%

Low extractables will reduce the stickiness and blocking of films. This is particularly significant in markets for soft films and food contact applications. The comparative film blocking forces of two LLDPE grades of 0.920 g/cm^3 density are:

- mLLDPE blocking force = <0.1 g
- LLDPE blocking force = 5 g

A typical example of the improved film impact strengths of 0.920 g/cm^3 density LLDPE produced by metallocene and Z-N catalysts is:

- mLLDPE = 1000 g
- LLDPE = 300 g

On the down side, metallocene polyethylenes suffer from rheology related deficiencies. Solutions are possible similar to those used with LLDPE to eliminate sharkskin and reduce pressure in film processes:

- Low shear sensitivity
- Low melt strength (tension thinning)
- Melt fracture (mainly in blown film processes)

One of the challenges is to develop molecular architectures that will mimic high pressure LDPE. The aim is to induce long chain branching in the correct amount and sequence in order to increase melt strength and minimize sharkskin formation in film processes.

Over the years polymer producers have invested considerable effort to improve the processability of LLDPEs for extrusion coating and tubular film processes. Constrained geometry metallocene catalysts that can induce long chain branching have been developed. **Table 13.11** shows the effects of such modifications on a couple of examples (#1 and #2). Although these resins have higher melt strength than the conventional metallocene PE, these are still substantially lower than that achieved by a typical high pressure LDPE. The claim is that better results are obtained in terms of lower neck-in and higher line speeds, without draw resonance, compared to conventional linears. Other approaches have used metallocene catalysis in dual reactor trains to induce long chain branching and broaden molecular weight distribution.

Table 13.11

Properties of mPEs

Polymer	Melt index (dg/min.)	Density (g/cm^3)	Melt strength (cN)	Mw/Mn ratio
Resin (C_8) #1	4.75	0.91	1.3	2.11
Resin (C_8) #2	6	0.90	1.27	2.18
mLLDPE (C_6)	10	0.905	0.349	2.32
LDPE (autoclave)	0.42	0.918	25.4	7.6

Some patents claim the use of two α-olefins in a dual reactor processes. Others combine slurry with gas phase reactors in cascade. The claims are that the molecular architecture achieved increases melt strength and improves processability. Most of these claims do show some increased melt strength, but it is not clear whether the extensional viscosity remains tension thinning or whether it is transformed into the tension hardening response typical of LDPE.

Metallocene polymerization catalysts technologies have also been extended to the production of polypropylene homo and copolymers. These have narrow MWD (<2.5) and improved crystallization behaviour and reduced crystallite size. Interesting materials with compositions that contain around 80% isotactic

PP with the balance made up of ethylene and other α-olefins have been commercialized. These products can be blended with other polyolefins (PE/PP) to enhance properties such as adhesion and puncture resistance.

Chapter 14

Barrier and other polymers

Traditionally the extrusion coater and laminator have depended on aluminium foil to provide a barrier to gases, moisture and light. Structures combining paper, aluminium and polyethylene have an outstanding combination of attributes at an attractive cost. Their key attributes are listed in **Table 14.1**. Tandem extrusion coating is probably the ideal process to combine these materials in a continuous high speed operation.

Table 14.1

Attributes of paper/foil/PE laminates

Positive attributes
Excellent gas, light and moisture barrier
Stiffness and strength from paper or cardboard
Sealability from PE coating
Moisture protection from PE coating
Heat sealing with HF induction heating
Light and low cost materials
Negative attributes
Poor flex. crack resistance of low gauge Al foil
Attacked by acid foods (pH <6.4)
Recyclability difficult
Opaque/not transparent
Not microwaveable

The negative items on the list can be overcome by replacing aluminium with a barrier polymer that is either coextrusion coated with polyethylene or film laminated. Polymers with high gas barrier are, however, costly and have a number of deficiencies.

Unlike metal and glass, plastic barrier materials are permeable to small molecules. **Figure 14.1** shows the oxygen and moisture transmission rates of the key polymers used in flexible packaging applications. The highly polar polymers like EVOH and polyamide have good barrier to oxygen, but are poor with water vapour. Conversely the non-polar polyolefins are good in resisting moisture, but poor with oxygen. It should also be borne in mind that unlike metal or glass, these polymers will undergo changes in permeability as a function of humidity and temperature. During heat treatment processes such as pasteurisation (80-90 °C) and sterilization (121 °C), gas permeability will increase.

The two best commercially available barrier resins are ethylene-vinyl alcohol (EVOH) and polyvinylidene chloride copolymers (PVDC). However, because of the preponderance of hydroxyl groups (-OH) in its backbone, EVOH is very moisture sensitive. Overall PVDC has the better balance of properties since it combines good gas barrier with excellent resistance to moisture.

The disadvantages of PVDC are twofold. Firstly the polymer's processing temperature is very near its degradation temperature (160-170 °C) at which point hydrochloric acid (HCl) can be formed. The polymer must be processed on machines that are corrosion resistant, use low shear screws and have very precise temperature controllers. The second is the trend to eliminate the use of chlorinated materials for environmental reasons. This has tended to favour EVOH in spite of its deficiencies and high cost.

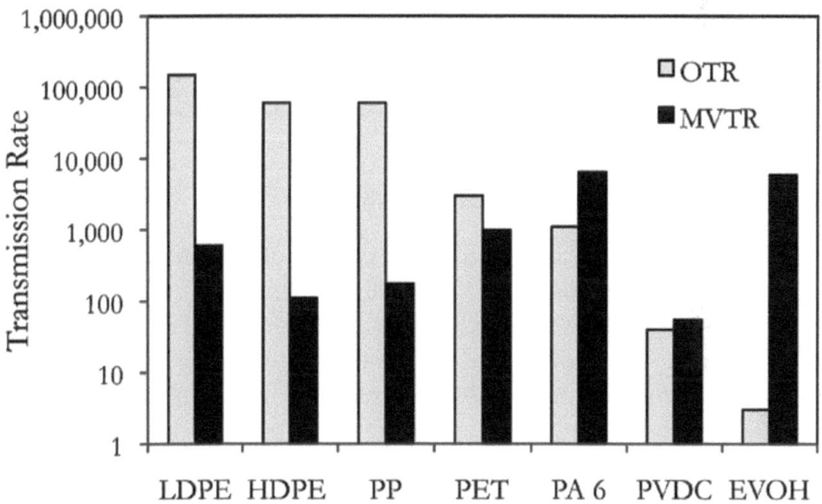

Figure 14.1

Comparative oxygen and moisture barrier properties of various packaging polymers
(OTR : $cm^3.\mu m/m^2$.day.atm. @ 23 °C, 50% RH
MVTR: $g.\mu m/m^2$.day @ 38 °C, 90% RH)

PVDC resins can contain up to 18% vinyl chloride as comonomer to improve processability. Some grades can also include methyl acrylate and acrylonitrile comonomers in the formulation. Typical processing aids and stabilizers are epoxidized soya bean oil and tetra-sodium pyrophosphate.

Many of the PVDC application processes are in liquid form. These resins do not contain plasticizer and are usually applied with gravure roll coating techniques using aqueous (latex) or solvent based diluents. Multiple passes are required to build-up sufficient coating weight. Drying is carried out in air heated tunnels. To minimize coating weight on porous paper supports, a primer of LDPE can be extrusion coated on the paper, corona treated and passed through the PVDC roller applicator and dried. BOPP, OPET and OPA films coated with PVDC are widely used in snack food and confectionery packaging. The PVDC provides barrier and sealability to these oriented films.

Coextruded film and sheet from flat die processes based on a PVDC barrier have been commercially used for many years. The PVDC is encapsulated with an EVA and sandwiched in between two layers of polystyrene (HIPS) and/or polyethylene. HDPE can be used on the inner side to improve product resistance. The sheet is thermoformed to produce plastic cans and other deep draw containers. EVAs with 26-28% VA have excellent adhesion to PVDC, polystyrene and polyolefins. Other adhesive resins, such as hydrogenated styrene butadiene block copolymers (SEBS) are also cited.

14.1 Ethylene-vinyl alcohol polymers (EVOH)

EVOH polymer is expensive to produce as it cannot be polymerized from its monomers. The manufacturing process involves two steps. The first is to produce an ethylene-vinyl acetate (EVA) copolymer containing 60-70 mol percent vinyl acetate. The EVA is then hydrolyzed or saponified into the alcohol. This reaction is slow and produces more than 30% of unwanted by-product in the form of acetate salts. This weight loss is due to the stoichiometric chemical reaction, which converts the ester groups (-$COOCH_3$) to the much lighter hydroxyls (-OH). This by-product is of low value and the processes to regenerate acetic acid for recycling into vinyl acetate monomer are cumbersome.

Although more heat stable than the PVDC polymers, EVOH should not be extruded above 250 °C. At higher temperatures the polymer will begin to degrade forming gels, which are very difficult to purge out of the machine. This temperature ceiling limits its use in the extrusion coating process unless low temperature adhesion promotion methods are used. EVOH will normally be used in a coextruded or laminated structure with a polyolefin to protect it from moisture. These highly polar materials are incompatible with polyolefins and require specially formulated tie-resins for interlayer adhesion. These are normally based on maleic anhydride modified polyolefins and are also costly. These modified polymers can also be used as compatibilizers to recycle scrap coextruded EVOH films.

The chemical structure of EVOH is shown below. The hydroxyl group is hydrophilic and the polymer must always be kept sealed in moisture-proof packaging.

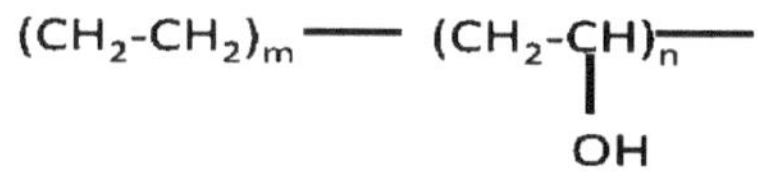

Hydroxyl

The excellent gas and aroma barrier of EVOH polymers are due to their high crystallinity and the high level of hydrogen bonding induced by the hydroxyl groups. These materials are defined by their ethylene content, in mol%, and by a measure of their melt flow rate. The melt flow index is usually measured at 210 °C with a 2.16 kg load. Ethylene content determines the gas permeability of the polymer. Increasing ethylene incorporation will reduce moisture sensitivity, but will increase the oxygen transmission rate. High ethylene will also improve processability and thermal stability. Thus a trade-off has to be made by the converter in deciding whether some oxygen barrier can be sacrificed for reduced sensitivity to humidity and easier converting. The effect of four levels of ethylene on oxygen permeability (OTR) is shown in **Table 14.2**.

Table 14.2

Effect of ethylene content on OTR

Test	Ethylene mol.%			
25 µm OTR cm^3/m^2.day atm.	**29**	**32**	**38**	**44**
0% RH @ 25 °C	0.2	0.4	0.5	1.5
75% RH @ 25 °C	0.8	0.9	1.4	2.6

EVOH grades designed for coextrusion coating applications are available with 35 mol% ethylene. These 8 dg/min melt index resins can apparently be processed at 280 °C. 25 µm OTR values are 0.55 cm^3/day @ 25 °C, 0% RH. **Table 14.3** compares three EVOH grades. The melting point drops as ethylene content increases. More recently grades with melt indices up to 16 dg/min. have been commercialized.

Table 14.3

Properties of various EVOH grades with different ethylene content

Property	Method / value	A	B	C
Mol.% ethylene	internal	32	35	44
Density (g/cm^3)	ASTM	1.19	1.18	1.14
Melt index (dg/min)	190 °C	4.4	8.0	5.5
	210 °C	10.5	16	13
Melting point (°C)	DSC	183	179	165

High ethylene content EVOH is claimed to offer the following benefits:

- High speed coating with low motor load
- Consistent coating thickness down to 1 μm is achieved
- Excellent gas barrier and aroma and flavour retention
- Excellent thermal stability at these high operating temperatures (no voids, gels, etc.)

Processing guidelines based on pilot plant trials have been proposed for the following flexible packaging structure:

12 μm OPET
5 μm Adhesive tie-layer
EVOH
5 μm Adhesive tie-layer
40 μm PE film

The 40 μm PE film is extrusion laminated in situ with the 3 layer coextrusion onto the OPET film substrate. The adhesive polymer has a melt index of 5.6 dg/min. It is recommended to raise the adhesive temperature to 300 °C.

The recommended EVOH barrel temperature settings are: 175/220/280/280 °C. The feedblock and die temperatures were also set at 280 °C. The EVOH thickness was varied from 1.5 to 10 μm at a line speed of 150 mpm. Neck-in width was 45 mm.

The products that could be of interest to the extrusion coater are polyethylene coextruded or laminated with EVOH and applied on paper and board substrates (see section 2.5).

EVOH based laminates can be utilized in microwaveable applications where aluminium containing structures are unsuitable. Another application is in the partial replacement of LDPE in fruit juice packaging. LDPE in **direct** contact with juices will scalp, i.e., absorb flavour components and vitamins thus compromising the taste and nutritional quality of the produce. The highly polar barrier polymers are much less susceptible to this scalping phenomenon and can be used to reduce these losses. A further advantage is that EVOH has better resistance to acid foods than aluminium.

The orientation of coextruded EVOH films offers the possibility of much improved properties. Coextrusions with EVA, LD/LLDPE and polyamide can be oriented by either double bubble or tenter frame processes. Barrier and moisture resistance are improved by the orientation process. Polyamide can be coextruded with EVOH without a tie-layer. This combination with a polyolefin seal layer offers excellent gas and aroma barrier. These films will also have excellent puncture and flex resistance.

For more information on processing of EVOH see sections 3.7 and 4.4.

14.2 Polypropylene in extrusion coating

Extrusion coating of polypropylene has been of interest to coaters for many years. Commercially available PP develops melt resonance at line speeds above 70 mpm. To overcome this limitation the polymer is compounded with coating grade LDPE (15-20%). However, unlike HDPE, PP is difficult to blend with branched polyethylene and requires an additional melt homogenization step. There are some patents that describe the use of compatibilizers such as ethylene propylene elastomer to improve miscibility. There is also some evidence that MAH grafted EPDM enhances miscibility. These melt blends tend to be costly and are only used in applications that can accept the added costs.

New technologies based on metallocenes have enabled the production of polymers based on propylene with ethylene and other α-olefins. These products contain approximately 80% isotactic polypropylene, and have enhanced adhesion and toughness. They are claimed to be excellent modifiers for PE and PP.

An important limitation in the use of PP in extrusion coating is its intrinsically poor adhesion to non-porous substrates. Unlike LDPE it does not form sufficient oxidized surface molecules in the air gap to promote adhesion to the substrate. At high extrusion temperature PP will degrade with catastrophic chain scission. To overcome this degradation high levels of stabilisers are added, which in turn limit oxidation in the air gap. Limited adhesion, poor heat stability and draw resonance at modest line speed have discouraged the exploitation of this polymer by extrusion coaters.

Some polymer suppliers offer PP materials tailored for extrusion coating. Grades capable of running up to 200 mpm with neck-in values of 70 mm per edge at melt temperatures around 310 °C are claimed to have been commercialized. Coating weights of 20 g/m^2 are achieved. Draw resonance is apparently avoided at these conditions. Good adhesion, however, is only possible on porous substrates and polypropylene fabrics, such as raffia.

Isotactic PP with long chain branching has been developed. The polymer is subjected to a proprietary treatment, which cleaves polymer chains and recombines the fragments as long chains on the PP backbone. This product is claimed to be strain hardening and has been successfully utilized in extrusion coating. Variants of these products, which are miscible with polypropylene, are available and suitable for extrusion coating.

PP offers a number of property advantages over LDPE:

- Improved grease resistance
- Heat resistance: melting point of 160-165 °C.
- Improved **non-stick** surfaces for packaging, e.g. frozen fish and meat
- High gloss coatings
- Excellent chill roll release
- High stiffness
- The ability to heat sterilize PP films suggests uses in medical applications

The largest current application for coating grade PP is on woven PP tape fabrics (raffia). The extruded PP is expected to adhere to these fabrics with short air gaps. These materials are used as heavy duty sacks for fertilizers and tarpaulins in markets such as India. UV stabilizers are added for outdoor applications.

A potentially large market area for PP coated board is in microwaveable trays. In this arena PET (polyester) is the principle competitor. PET has the advantage that it will withstand normal oven conditions as well as microwaving. However PP suppliers highlight the advantageous properties of PP presented in **Table 14.4**.

Comparative processing data is provided in section 4.4.

Table 14.4
Comparison of PP-coated board and PET for microwaveable packaging

Property	PP (blend)	PET
Density (kg/m^3)	910	1400
Yield (25 μm coating)	44 m^2/kg	30 m^2/kg
Neck-in (mm)	60	130-150
Surging	None	High
Pre-drying	None	Yes

14.3 Polyester (PET)

Polyesters are of interest for specialized applications where exceptional heat resistance, stiffness and solvent resistance are required. The best known applications are coated paperboard for ovenable trays and coated papers for electrical cable wrap. The patent literature teaches methods of coating specially formulated polyester resins for extrusion coating on aluminium or steel for the manufacture of cans using the draw-redraw process.

An important advantage that PET has over polyolefins is in its very low sorption (scalping) of nutrients, flavour, vitamins and aroma compounds found in food and cosmetic products. The rate of sorption of these compounds by PET is at least one order of magnitude lower than polyethylene.

Polyethylene terephthalate (PET) is produced by the esterification of ethylene glycol with a dibasic acid. Terephthalic acid (produced by the oxidation of paraxylene) or dimethyl terephthalate is reacted with the glycol to form diethylene glycol terephthalate, which is the monomer for PET. Polycondensation of the monomer takes place in vacuum reactors designed to evaporate ethylene glycol to shift the reaction equilibrium towards long polymer chains. The basic chemical formula for PET is shown below:

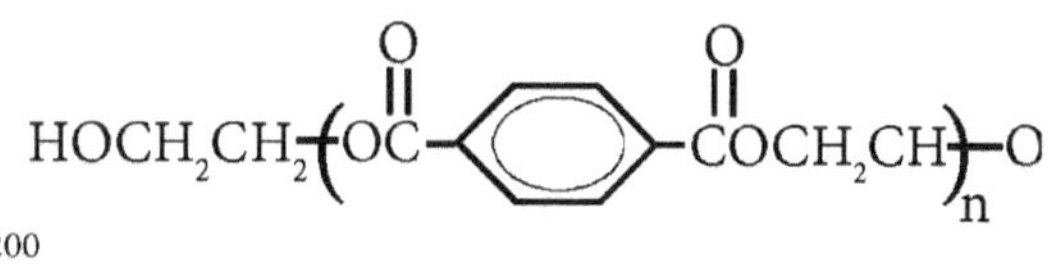

n = 100-200

POLYETHYLENE TEREPHTHALATE

Copolyesters are made by using more than one glycol and/or dibasic acid in the reaction. This reduces crystallization and melting temperature and forms a clear amorphous film or sheet. The copolyesters remove some of the processing restrictions. The melting point is reduced from 260 to 220 °C by forming the copolyester. These products referred to as PETG (glycol modified) are clear amorphous materials.

It is important to avoid hydrolytic degradation when extruding PET resins. Degradation leads to discoloration, chain scissions, crosslinking and the formation of acetaldehyde. The polymer must, therefore, be free of trace moisture when extruded. Moisture will also lead to melt instabilities in the air gap.

The properties of polyesters are very dependent on their state of crystallization. In the coating process they are extruded at 290-300 °C, which is well above their glass transition temperatures (Tg). Consequently they will form amorphous coatings. In the amorphous state PET is clear and heat sealable and can be thermoformed. In the crystalline state PET is opaque, stiffer and much more difficult to heat seal.

Polyesters recommended for extrusion applications typically have the following intrinsic viscosity values:

- OPET films: 0.60-0.70 dl/g
- Extrusion coating grades: 0.70-1.00 dl/g

The intrinsic viscosity of polyester is measured in decilitres per gram (dℓ/g) and is dependent upon the length of its polymer chains. The longer the polymer chains, the more entanglements between chains, therefore, the higher the viscosity.

High viscosity PET copolymers are recommended for extrusion coating applications. These are dried in-line with hot air at a temperature of 60-65 °C to guarantee moisture content below 100 ppm. PET resins are available which are processable in extruders fitted with screws designed for polyethylene. Neck-in will be around double that experienced with LDPE. Fibre tear adhesion to paper substrates at coating weights down to 25 g/m^2 at a line speed of 150 mpm is claimed.

Preheating the paper to 120-150 °C is suggested. Flame or IR heaters can be used for preheating. However, heating cardboard for hot ovenable trays will reduce moisture content causing the board to possibly tear during die pressing. An alternative is to use corona treatment, which has been shown to substantially improve adhesion to paper and paperboards. Fibre tearing adhesion is claimed to board used for ovenable trays.

To reduce coating weight on paper substrates it is recommended to first coat an EVA (12-18% VA) layer as primer followed by the polyester layer. This will enable a significant lowering of the polyester coating weight.

When coated onto paperboard in the amorphous state polyesters will crystallize when exposed to the high temperatures used during oven heating and can, therefore, be used in dual ovenable food applications. These PET coated boards can resist temperatures of 200 °C.

Amorphous (APET) polyester resins have been developed for extrusion coating. These are used as sealant layers to seal to glycol-modified polyesters (PETG) for improved processability, to crystallisable PET (CPET) widely used for ovenable thermoformed trays, and to PVC and polycarbonate sheet. Peelable steam sterilizable seals can be produced. These materials will require higher temperatures (~200 °C) and longer dwell times than polyolefins for effective seals to be formed.

For lidding stock OPET films coated or laminated with APET are used for packaging refrigerated and frozen foods. The films will resist the temperature range from -40 to 205 °C.

In addition to high temperature resistance APET sealants have excellent resistance to chemicals, oxidizing agents, and oils. Extractables are also very low.

Recommended further reading

Oveby C. Rheological Indicators of LDPE to Predict Processing Performance in Extrusion Coating. Annual Transactions of the Nordic Rheology Society, vol. 13, **2005**

Massouda D. F., U.S. Patent 5116649, **1992**

Morris, B.A. and Suzuki, N., The Case Against Oxidation as the Primary Factor For Bonding Acid Copolymers to Foil, ANTEC 2001, SPE, **2001**, p. 25-35

Antonov, V. and Soutar, A.M., Foil Adhesion with Copolymers: Time in the Air Gap, TAPPI Polymers, Laminations and Coatings Conference, Atlanta, TAPPI Press, **1991**, p.553

Morris, B.A., The Influence of Stress on Peel Strength of Acid Copolymers to Foil, ANTEC 1999, SPE, **1999**, p. 34

Cheney G., Markgraf D., Benson M., Statistical Analysis of the effects of Ozone on Adhesion in the Extrusion Coating Process, TAPPI Polymers, Laminations and Coatings Conference, Toronto, TAPPI Press, **1997**, p. 649-655

Greig S., Jadon N., Sherman Treaters Ltd. Corona, Ozone and Flame Treaters for Extrusion Coating Lines, TAPPI European PLC Conference, TAPPI Press, **2001**

van Krevelen, D.W. and te Nijenhuis, K., Properties of Polymers, 4th Edition, **2009**, Elsevier Science, 9780080548197, Ch. 7

Karagiannis, A., Mavridis, H., Hrymak, A.N. and Vlachopoulos, J., Polymer Engineering & Science, **1988**, 28, p. 982–988

Zisman, W.A., Contact Angle, Wettability and Adhesion, ACS Advances in Chemistry series, **1964**, 43, p, 1-51

Laiho, E., Vaehaelae, M. and Sainio, M., USPTO patent application 20080261064

Kouda, S., Prediction of Processability in Extrusion Coating for Low Density Polyethylene, Polymer Engineering and Science, **2008**, 48, p. 1094–1102

Tracton, A.A., ed., Coatings Technology Handbook, 3rd Edition, **2005**, CRC Press, 9781574446494

Kollbach, G., Sauer, R., Polyurethane Lamination Adhesive. US Patent 20100010156, **2010**

Schrenk, W.J., Bradley, N.L., Alfrey Jr., T., Maack, H., Interfacial Flow Instability in Coextrusion, Polymer Engineering and Science, **1978**, 18, p. 620-623

Dooley J., Determining the Processability of Coextruded Structures, TAPPI Polymers, Laminations and Coatings Conference, St. Louis, **2007**, TAPPI Press, p. 480

Helmy H., Recent Developments in Flat Film and Sheet Dies, Plastic Film and Sheeting, **1988**, 4, p. 193

Rauwendaal, C., Understanding Extrusion 2nd Edition, **2010**, Hanser Verlag, Munich, 9781569904534

Fischer, P., Wortberg, J., New Spiral Die Designs, Innovations in Extrusion Conference, SKZ, **1998**

Shelat , K.J., Dutta N.K., Choudhury, N.R., Interfacial interaction and Morphology of EVOH and Ionomer blends using Scanning Thermal Microscopy and its corelation with barrier characteristics, Langmuir, **2008**, 24, p. 5464

Fetell, A.I., Nylon-Containing Ionomer Improves the Performance of EVOH Copolymers in Multilayer Structures, TAPPI Polymers, Laminations and Coatings Conference, TAPPI Press, **1997**, p. 503

Sacharow, S., Griffin, R.C., Principles of Food Packaging 2nd Edition, **1980**, AVI Publishing, 9780870553479

Constant, D.R., Basic Polymer Rheology as Related to Extrusion Coating Machinery. 11th TAPPI European PLACE Conference, Athens, TAPPI Press, **2007**

Graeff, R.W., Getting Optimum Results from PET Drying, Mod. Plastics Intl, May **1992**, p. 99

Pechinski, L.J., Changing Demands Spur Adhesive Developments, Paper Film and Foil Converter, **1991**

Bezigian, T., The Effect of Corona Discharge on Polymer Films, TAPPI Journal, March **1992**, 75, p. 139-141

Cogswell, F. N., Polymer Melt Rheology During Elongational Flow, ACS Symposium, Philadelphia, **1975**

Brodie, V., Taste Testing as it Relates to Packaging Resins, TAPPI Journal, Dec. **1988**

Abad, M.J., et al, Use of a sodium ionomer as compatibilizer in polypropylene-barrier EVOH copolymer blends, Journal of Applied Polymer Science, **2004**, 94, p. 1763-1770

Agalloco, J., Akers, J., Madsen, R., Aseptic Processing: A Review of Current Industry Practice, Pharmaceutical Technology, October **2004**, p. 126

Index

C

N

O

P

U

V

W

Z

FSC
www.fsc.org
MIX
Papier | Fördert gute Waldnutzung
FSC® C083411

Kolibri 360 GmbH
Ferdinand-Jühlke-Straße 7
99095 Erfurt, Deutschland
produktsicherheit@kolibri360.de